FOR RANDY

What a man needs in gardening
is a cast-iron back, with a hinge in it.

CHARLES DUDLEY WARNER, 1870

ACKNOWLEDGMENTS

I would like to thank all of the many people who provided encouragement, advice, and information during the preparation of this book. Greatly helpful were the University of Massachusetts Library (Amherst, Massachusetts), the Petersham Public Library (Petersham, Massachusetts), the Chester County Library (Exton, Pennsylvania), and the Salem Free Public Library (Salem, New Jersey). The staffs of the Massachusetts Horticultural Society (Boston, Massachusetts), the National Potato Council (Denver, Colorado), the Corn Refiners Association, Inc. (Washington, D.C.), and the Popcorn Institute (Chicago, Illinois) all supplied valuable information, as did the International Chili Society (Newport Beach, California), Caliente Chili, Inc. (Austin, Texas), DNA Plant Technology Corporation (Cinnaminson, New Jersey), Molecular Genetics, Inc. (Minnetonka, Minnesota), and Twyford Plant Laboratories, Inc. (Santa Paula, California). I am especially grateful for assistance given by Jeanette Lowe of W. Atlee Burpee Company (Warminster, Pennsylvania), Corinne Willard of Comstock, Ferre & Co. (Wethersfield, Connecticut), Ron Driskill of Siberia Seeds (Olds, Alberta, Canada), the staff of the Heirloom Garden Project, Vegetable Crops Department, Cornell University, and Rob Donaldson of George Washington University, a master at answering difficult questions.

I wish particularly to thank Tom Rawls, without whose encouragement this book would never have been written, and my editors, Jill Mason and Deborah Burns. Finally I would like to thank Harry Buell, who plowed our garden, my husband, Randy, who bought me a fountain pen for Christmas, and my grandmother, Mildred Bliss, who taught me how to read.

CONTENTS

VEGETABLES IN AND OUT OF THE GARDEN

Nineteenth-century gardener Dean Hole, dwelling cheerfully on the subject of his craft in 1899, wrote:

I asked a schoolboy, in the sweet summertime, "What he thought a garden was for?" and he said, *Strawberries.* His younger sister suggested *Croquet,* and the elder *Garden-parties.* The brother from Oxford made a prompt declaration in favor of *Lawn Tennis* and *Cigarettes,* but he was rebuked by a solemn senior, who wore spectacles, and more back hair than is usual with males, and was told that "a garden was designed for botanical research, and for the classification of plants." He was about to demonstrate the differences between *Acoty-* and *Monocoty-ledonous* divisions, when the collegian remembered an engagement elsewhere.

Nowadays gardeners have similarly scattered and unscientific options: according to a 1985 poll sponsored by the National Gardening Association in Burlington, Vermont, 30 percent of respondents claim their gardens are for fresh vegetables, and another 25 percent state theirs are for better-tasting, higher-quality food. A light-hearted 22 percent—the croquet and lawn tennis bunch—garden for fun, and a serious-minded 15 percent garden to save money. The remaining 8 percent presumably garden for quirkily unclassifiable reasons: a quick poll of our neighborhood, besides the standard percentage of food/fun answers, turned up one misanthrope who plants for peace and quiet, one hypertensive who plants for medically prescribed relaxation, and one professor who likes the feel of sun on his bald spot. All told, such gardeners make up 40 percent of American households, a grand total of some thirty-three million home vegetable gardens. Top vegetable in these gardens, by a considerable margin, is the tomato, followed in dwindling order by peppers, green beans, cucumbers, onions, lettuce, squash, carrots, radishes, and sweet corn. National home production, according to one estimate, totals sixteen billion dollars' worth of food annually, a substantial economic contribution at a time when every stray billion counts.

Vegetables and fruits, on the average, make up about 10 percent of our average daily caloric intake, contributing an unimpressive 1 percent of our daily fat and 7 percent of our daily protein. Their major importance is as sources of vitamins and minerals: fruits and vegetables provide 90 percent of our vitamin C, 50 percent vitamin A, 35 percent vitamin B_6, 20 percent each niacin (B_3) and thiamine (B_1), 20 percent iron, and 25 percent magnesium. Dedicated vegetarians, to maintain a nutritiously balanced diet, combine vegetables and fruits with cereals, eggs, nuts, and dairy products. Such dietary juggling is necessary because plant-derived proteins, sadly, are incomplete, lacking one or more of the essential amino acids required for human health. Beans, for example, are generally low in methionine; corn and other cereal grains lack lysine. Meat, on the other hand, is complete protein: the Neanderthals subsisted nicely on it and Methuselah's steady diet of it is said to have supported a lifespan of 969 years.

Vegetables—"rude herbs and roots"—have a long history of disdainful neglect, with the exception perhaps of the ubiquitous onion.

While meat-eating has its nutritional advantages, vegetable-eating traditionally has been considered superior for the health of the spirit. Ethically minded human beings have practiced vegetarianism for thousands of years. Among its earliest proponents was the Greek poet Hesiod, who preached—but probably did not practice—a romantic rural diet of "mallow and asphodel." (He also urged his readers to get a good woman, for plowing.) He was followed—after a two-hundred-year lag—by Pythagoras, sometimes called "the Father of Vegetarianism," who urged a meatless diet upon his disciples in the sixth century B.C. Such a diet was thought to promote peace of mind and to suppress distracting animal passions—common aims among budding philosophers—and, since the Pythagoreans believed in the transmigration of souls, to avoid the sinful slaughter of fellow spirits. The ban on animal foods was extended to fava beans, also possible repositories of human souls, and one story holds that a group of embattled Pythagoreans, caught between a rock and a hard place, allowed themselves to be slaughtered rather than escape by trampling through a bean field.

Europeans of the belligerent Middle Ages much preferred to eat meat, unencumbered by spiritually superior vegetables. European vegetables—"rude herbs and roots"—have a long history of disdainful neglect, with the exception perhaps of the ubiquitous onion. While the medieval peasants made do with grain porridges and soups, their social superiors dined on swan, crane, and peacock, often complete with feathers; roast piglet draped in daffodils; chicken doused in almond milk; meatballs wrapped in gold foil; venison; and veal. "Kitchen" gardens of the time reflected the prevailing carnivorous atmosphere: little but onions, leeks, garlic, and cabbages was grown, and, as field crops, peas and beans. In some instances, the gardens of the wealthy grew no rude herbs at all, but were turned over to useful inedibles such as flax and hemp.

Vegetable consciousness was generally slow to arise, and, in the sixteenth century, when new plant foods—tomatoes, potatoes, peppers, pumpkins, sweet corn, Jerusalem artichokes, "French" beans—began pouring in from the New World, they were received with gravest suspicion. Modern pro-vegetable writers find this wholly unreasonable. Cyril Connolly writes: "Never would it occur to a child that sheep, pigs, cows or chickens were good to eat, while, like Milton's *Adam,* he would readily make a meal off fruit, nuts, thyme, mint, peas, and broad beans. . . ." While it's difficult to predict just how a hungry six-year-old would view hamburger on the hoof, as a general rule of thumb people are neophobic at the dinner table—that is, reluctant to try the unfamiliar. This distrustful avoidance behavior, coupled with its equally powerful opposite, incurable inquisitiveness about new foods, is referred to by food experts Elizabeth and Paul Rozin as the "omnivore's dilemma." Omnivores, or food generalists, who have a widely varied diet—as opposed to food specialists like giant pandas, which live exclusively on bamboo—must continually weigh the risks of possible poisoning against the benefits of discovering a new, nutritious, and delicious edible.

This omnivorous struggle between caution and curiosity has shaped the course of human history. In the most radical instances of neophobia, well-meaning attempts to relieve famines with shipments of culturally off-the-wall foods have failed abysmally. In 1800, starving Italians rejected a boatload of alien Irish potatoes; in the 1840s, the famished Irish spurned a donation of imported Indian corn. In contrast, people, insatiably experimental, can—and do—learn to love practically anything, which explains why our omnivorous species has spawned such a vast number of culinary themes and traditions. Human beings, in various times and places, have revelled in slugs, grasshoppers, sea cucumbers, fermented fish sauces, flamingo tongues, fugu fish, and chile peppers, and Henry David Thoreau, defending human adaptability, claimed he could live on tenpenny nails. He doesn't seem to have tried, but during his stint at Walden Pond he did sample roasted woodchuck and fried rat.

In the eighteenth century, diets began to diversify and the creative idea of the all-vegetable regime began to resurface, though the formal term *vegetarian* was not to appear in print until the founding of the British Vegetarian Society—popularly considered a bunch of oddballs—in 1847. Eighteenth-century vegetarianism constituted a dietary about-face. Scholarly publications to date had warned diners off vegetables on the grounds that they caused "wind"—an obsessive social concern—and melancholy. (Fresh fruits were suspected sources of fevers: Samuel Pepys, in his diary entry of August 12, 1661, worries guiltily that the fruit he gave Lord Hinchingbroke may have given him smallpox.) In colonial America, one of the earliest vegetable advocates was Benjamin Franklin, who bravely experimented with a vegetarian diet in his youthful days as a Philadelphia printer, in company with Samuel Keimer, his employer. They tried it for three months, and Franklin later recalled:

I went on pleasantly, but poor Keimer suffered grievously, tired of the project, long'd for the flesh-pots of Egypt, and order'd roast pig. He invited me and two women friends to dine with him; but, it being brought too soon upon the table, he could not resist the temptation and ate the whole before we came. . . .

Franklin himself held on longer, but finally fell off the wagon on a sea voyage from Boston. The boat became stranded off Block Island, vegetarian victuals were in short supply, and a hungry Franklin gave in to an irresistible meal of fresh-caught fried codfish. Despite his fall from dietary grace, Franklin continued to advocate vegetarianism, and his popular *Poor Richard's Almanack* (1732) is peppered with references to the benefits of abstinence, as in "Cheese and salt meat should be sparingly eat" and "Hunger is the best Pickle."

Franklin's vegetarian views derived largely from health and economic motives, plus a humanitarian repugnance for the "unprovoked murder" of animals. Pro-vegetable contemporaries, however, often argued their case from the you-are-what-you-eat standpoint: that is, those who eat animals act like animals. Prominent among these was the French philosopher Jean-Jacques Rousseau, who in his educational treatise *Emile* (1762) used dietary practice to take a political poke at the offensive English: "It is a fact that great eaters of meat are in general more cruel and ferocious than other men; this observation holds good in all places and at all times; the barbarism of the English is well known." Rousseau, however, preached more than practiced, and in daily life happily ate everything but asparagus, which bothered his bladder.

Franklin himself held on longer, but finally off the wagon on a sea voyage from Boston. The boat became stranded off Block Island, vegetarian victuals were in short supply, and a hungry Franklin gave in to a meal of fresh-caught fried codfish.

Nonetheless meat continued to dominate most menus. Nineteenth-century meals were enlivened only by turnips, potatoes, and pickles; and the English visitor Frances Trollope, touring the United States in the 1830s, commented in appalled tones upon the farmers of Maryland, who had no vegetable gardens, but subsisted dully on pork, salt fish, and cornbread. Perhaps in reaction to such a cuisine, food faddism proliferated in the nineteenth century. The health-conscious struggled variously to forego alcohol, tea, coffee, cocoa, tobacco, condiments, milk, sugar, and meat. Prospective vegetarians were inspired by the example of romantic poet Percy Bysshe Shelley, whose persuasive anti-meat arguments were presented in "A Vindication of the Natural Diet" (1813). Shelley's vegetarianism was based on a belief in the wholesome spiritual effects of a plant-based diet: vegetable foods promoted "Health and Virtue," while animal foods generated "Disease, Superstition, and Crime." The natural diet as followed by the Shelley menage seemed to suit the poet himself, but found less favor with his wife, Harriet Westbrook, whose kindest comment was, "we do not find ourselves any the worse for it," and none at all with his son, Percy Florence Shelley, who claimed it made him hopelessly fat.

First among food faddists in the United States was the vociferous Sylvester Graham, whose name survives today in the ever-popular graham cracker. Graham was born in 1794 in West Suffield, Connecticut, youngest child in a family of seventeen. He was ordained a Presbyterian minister in 1826, but soon abandoned the cloth for the job of general agent with the Pennsylvania Society for the Suppression of Ardent Spirits, based in Philadelphia. During his tenure as temperance agent, Graham studied anatomy, physiology, and nutrition, and evolved the dietary theories that became the basis of the "Graham system." According to Graham, the "enormous wickedness and atrocious violence" preceding the Flood were the result of excessive meat-eating. Similarly, indulgence in meats, fats, salt, spices, ketchup, mustard, and Demon Rum was drastically weakening the present American populace, leaving citizens open to crime, sexual sin, and mental and physical disease. The cholera epidemic of 1832 seemed to prove his point and the chastened public seized upon dietary reform.

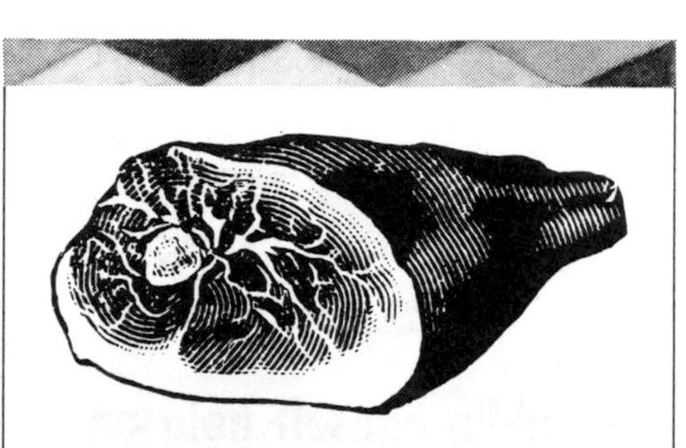

According to Graham, the "enormous wickedness and atrocious violence" preceding the Flood were the result of excessive meat-eating.

Reform was no laughing matter: the immensely popular Graham system encompassed an unappetizing diet of oatmeal porridge, beans, boiled rice, unbuttered whole-grain bread, and graham—originally Graham—crackers, plus cold baths, hard beds, open windows, and exercise. Despite opposition from physicians of the time, raw vegetables and fresh fruits were also on Graham's approved list, to be washed down with the recommended drink: pure cold water. Grahamite boarding houses dedicated to this masochistic program opened in New York City and Boston—residents rose by bell at 5 A.M. University students formed campus Graham clubs (by most accounts, small and short-lived); and the trustees of the Albany Orphan Asylum adopted the Graham system for their charges, a move which, according to one source, "aroused great controversy in the periodical press."

For all its Spartan peculiarities, Grahamism did improve the American diet, effectively furthering the cause of garden-fresh vegetables in a notably vitamin-deficient society. Salad greens, tomatoes, radishes, cauliflowers, asparagus, green beans, and spinach began to appear—though often sadly overboiled—on everyday tables. Full-blown vegetarianism was embraced by some, including such celebrities as Richard Wagner, Leo Tolstoy, and George Bernard Shaw, who abandoned meat at the age of twenty-five, announcing, "A man of my spiritual intensity does not eat corpses." Instead, he consumed zucchini au gratin, tomato-and-mushroom pie, stuffed eggplant, spaghetti, cucumber salad, curried chestnuts, artichoke soup, and strawberry ice cream.

Noticeably unseduced by such delectable vegetable lures were the athletes: where the Olympic runners ate onions and the Roman gladiators barley bread, nineteenth-century athletes favored red meat and dark beer. To pro-vegetable nutritionists, such a diet appeared annoyingly successful. Victor after victor triumphed on meat alone. In 1809, British athletic champion Robert Barclay walked one thousand miles in one thousand successive hours, nourished on nothing but heaping plates of beef and mutton. In a telling episode of the 1860s, the Oxford rowing crew trained exclusively on beef, beer, bread, and tea, with an occasional suppertime helping of watercress. The opposing Cambridge team ate unrestricted amounts of salad greens, potatoes, and fresh fruits. The period 1861–69 produced an unbroken succession of Oxford victories. Fueled by such incidents, the red-meat mystique persisted well into the twentieth century, when nutritional research found the high-carbohydrate meal—pasta and potatoes—to be a more efficient source of rapid energy.

Full-blown vegetarianism was embraced by some, including such celebrities as Richard Wagner, Leo Tolstoy, and George Bernard Shaw, who abandoned meat at the age of twenty-five announcing, "A man of my spiritual intensity does not eat corpses."

Recent research does support the healthiness— if not the moral superiority—of a vegetable-heavy diet. A 1985 study by scientists at the University of Munich found that the blood of omnivores—people who eat all foods, including meat—is significantly more viscous than that of strict vegetarians, while the blood viscosity of ovo-lacto-vegetarians, who add eggs and milk to their vegetable diets, falls somewhere between the two. Increased blood viscosity—a condition also brought on by such evil practices as smoking and avoidance of exercise—may indicate increased risk of heart disease. All in all, researchers feel, it's another point for the pro-vegetable crowd.

Not everybody, of course, *likes* vegetables. A French study of food dislikes conducted in the late

"... and worse than all is the Garden Seeds and Flower Seeds which you sold Mr. Wilks for me an Charged me £ 6.4s.2d Sterling. Not one of all the Seeds Came up Except the Asparrow Grass, So that my Garden is Lost for me this Year. P.S. The Tulip Roots you were pleased to make a present of to me are all Dead as well."

1960s found that nearly 60 percent of the respondents blacklisted vegetables, as opposed to a mere 4 percent fingering such classic dreadfuls as liver and tripe. Green vegetables were next-to-last on the list in a survey of food preferences among thirty-nine hundred males in the American military; ice cream, meat, and potatoes received middling ratings; and emphatic top billing went to beer. Despite such published antagonism, vegetable consumption in this country has been steadily increasing over the past century. The vast majority presently consumed are commercial products; home garden output, regrettably, isn't quite what it was in grandmother's day. The U.S. Department of Agriculture estimates that home gardens supplied 126 pounds of vegetables per person in 1910, as opposed to 50 pounds per person nowadays. The drop-off, according to the 1985 National Garden Association poll, may be due in part to increasing urbanization: 35 percent of non-gardeners attributed their plantlessness to lack of space. Another 28 percent pleaded not enough time, and a hammock-lolling 13 percent muttered about too much work.

"It is not graceful, and it makes one hot," wrote the enchanting Elizabeth, Countess von Arnim, busily planting her German garden in the 1890s, "but it is a blessed sort of work, and if Eve had had a spade in Paradise and known what to do with it, we should not have had all that sad business of the apple." Perhaps—but then a quick glance at the instructions that arrived with our limp-looking asparagus roots, which begin with an unpromising "Dig a trench two feet deep," is enough to make one toss in one's lot with the 13 percent. The history of gardening is an endless tale of perseverance in the face of such trials and tribulations. A letter of 1737 from Thomas Hancock—uncle of John, of the memorable signature—to his English seed dealer is horticulturally heart-rending: ". . . and worse than all is the Garden Seeds and Flower Seeds which you sold Mr. Wilks for me an Charged me £6.4s2d Sterling were not worth one farthing. Not one of all the Seeds Came up Except the Asparrow Grass, So that my Garden is Lost for me this Year." He ends up with a scathing "P.S. The Tulip Roots you were pleased to make a present of to me are all Dead as well." Abigail Adams writes to John in Philadelphia worrying about the effect of "Drougth" on the corn; Thomas Jefferson fails to grow horse beans and reports "poor success" with potatoes; Gilbert White, amateur naturalist and dedicated gardener, records the loss of potatoes, kidney beans, and nasturtiums to ice in the chilly June of 1787. No gardener, however, surrenders to the slings and arrows of outrageous agricultural fortune. Separated from her vegetable garden during her husband's first presidential term, Martha Washington writes home to Mount Vernon: "Impress it on the gardener to have everything in the garden that will be necessary in a house keeping way—as vegetables is the best part of our living in the country."

Gardens are inevitably a trade-off of successes and failures. Among our acquaintances are green-thumbed growers of Jerusalem artichokes and blue potatoes, tomatillos and cinnamon basil, strawberry popcorn and lemon cucumbers, chocolate peppers, rat-tailed radishes, and yard-long beans, but in our latest garden, the garlic chives never sprouted and deer ate the sweet corn. Still, the strawberries did well and we had a good crop of pumpkins ready for Halloween. Asked what they'd like to grow next year, our children pitched for carrots, doughnuts, and bluebirds, which is probably no worse than my craving for a perennial border, a trench-less asparagus bed, and a black walnut tree. Scraggy spots among current plantings, I find, are often overlaid in my mind's eye with next summer's successes, the superlative vegetables of the lush and leafy future.

"Half the interest of a garden," said Mrs. C. W. Earle, an understanding English gardener of the last century, "is in the constant exercise of the imagination."

TOMATOES

According to the U.S. Department of Agriculture, four out of five people prefer tomatoes to any other homegrown food. In fact, American tomato enthusiasm runs so high that garden-less fans have grown them in backyard barrels, patio pots, and window boxes, on the balconies of apartment buildings, on the decks of houseboats, and even, in the case of a few intrepid souls, on the roofs of Volkswagens. Over 90 percent of home gardeners put in tomato plants, and Ohio, a state with a firm grasp of priorities, has designated *Lycopersicon esculentum,* the garden tomato, as its official state fruit. It's hard to remember that many of our ancestors wouldn't have touched a tomato with a ten-foot pole.

The tomato came originally from western South America, where small-fruited wild forms, described by botanists as weedy and aggressive, still proliferate through Peru, Chile, Bolivia, and Ecuador, and north into Central America and Mexico. Europeans almost certainly picked up their first tomatoes in Mexico, though back home the new-found fruit was referred to misleadingly as *mala peruviana,* the apple of Peru. (A similar geographic confusion reigned over the turkey, a predominately North American bird, which under its own steam never made it farther south than Guatemala. The French, vague about exactly where Columbus had landed, called it the bird of India. The Germans and Dutch called it the Calcutta hen, and the English, who were eating it by the 1540s, plumped for turkey, which stuck.) The Peruvian apple, luckily managed to avoid such permanent linguistic inaccuracy, and the name *tomate,* or *tomata,* was soon popularly adopted, from the Nahuatl, or Aztec, word *tomatl.*

The invading Spaniards saw the *tomatl* growing in Montezuma's gardens in 1519 and described it recognizably, though in less than glowing terms: they found the sprawling vines scraggy and ugly. Still, Cortez brought tomato seeds back to Europe, along with the more spectacular plunder, and tomato plants were soon growing in the sunny gardens of Renaissance Spain. Early tomatoes were already remarkably similar to those grown today. Generations of selection in pre-Columbian Mexico had produced fruits in a range of sizes, shapes, and colors, among them orange, yellow, and white, as well as tomato red. These first fruits seem to have been ribbed or lobed, like a peeled orange, rather than round. Round fruits of the sort familiar to tomato-eaters today were not described until 1700, when French botanist Joseph Pitton de Tournefort finally mentioned a *"Lycopersicum rubro non striato."* The *non striato* means not lobed, or, by implication, round.

The Italian tomato was also known as *pomi dei Moro*, Moor's apple, which, legend holds, was misinterpreted by a lascivious visiting Frenchman as *pomme d'amour*, hence love-apple.

From Spain, the tomato is said to have travelled, via mysterious wandering Moors, to Morocco, and from Morocco, via Italian sailors, to Italy. It was well received in Italy, where it was called *pomo doro,* or golden apple, suggesting that yellow tomatoes were the first popularly available. Italian tomatoes were also known as *pomi dei Moro,* Moors' apples, which, legend holds, was misinterpreted by a lascivious visiting Frenchman as *pomme d'amour,* hence love-apple. Or the name may have developed naturally from the tomato's passionate reputation: the tomato, almost from its European introduction, was considered a sensationally effective aphrodisiac. In part, this may have been due to what sixteenth-century herbalists and physicians saw as an anatomical resemblance between the tomato and the human heart. According to the then-respected "Doctrine of Signatures," which ascribed specific medical uses to plants on the basis of their resemblance to specific human organs, this made the tomato a surefire bet as a love potion.

As aphrodisiacs went, however, the newly transplanted love-apple was quite mild. Centuries earlier, Pliny had suggested eating hippopotamus foot to increase sexual potency; Horace advised dried marrow and liver; and Petronius recommended a vast list of hopefuls, including pitch from the pomegranate tree, ass testicles, oysters, frogs, fava beans, onions, and snails' heads in sauce. Aristotle advocated peppermint oil, and, in 1657, herbalist William Coles classified artichokes, sea holly, potatoes, rocket, chocolate, and candied orchid roots as lust-provoking plants. To counteract the excitatory effects of the above, he recommended cannabis or hemlock, the last of which must certainly have done the trick: a lethal relative of the innocuous carrot, hemlock was the principal ingredient of the Athenian State Poison fatally drunk by Socrates.

Along with its aphrodisiacal aura, the tomato was believed to pack a poisonous punch of its own. The earliest known botanical description of the tomato is that of an Italian herbalist, Pietro Andrea Mattioli, who, in 1544, perceptively linked the tomato to a number of disreputable relatives, among them mandrake, henbane, and deadly nightshade. This association with known poisonous plants was distinctly off-putting and most likely the reason for the three-hundred-year hiatus before the tomato was accepted as an everyday article of diet. In fact, by mid-sixteenth century, the tomato was ominously nicknamed wolf peach ("peach" from its luscious appearance, "wolf" from its presumptive poisonous qualities) in analogy to pieces of aconite-sprinkled meat thrown out as bait to destroy wolves. This nickname, Latinized, has persisted as *Lycopersicon,* the modern scientific moniker for the tomato.

The poison rumor was not without its element of truth. Tomatoes, along with potatoes, tobacco, and petunias, belong to the nightshade family, Solonaceae. The toxicity of certain members of this family is attributable to chemical compounds called alkaloids, examples of which include morphine and quinine, along with the less attractive nicotine and strychnine. The major alkaloid in the tomato is tomatine, present only in the plant leaves and stems, and, to a lesser extent, in the green fruit, from which it disappears as the tomato ripens. Tomatine is considerably less potent than the lethal alkaloids of nightshade or belladonna, but it still doesn't pay to boil up a batch of tomato-vine stew. The boiled vines *can* be used to produce an attractive pale-yellow natural dye, and pre-pesticide nineteenth-century gardeners used a sprinkling of broth from cooked tomato foliage to destroy aphids.

For the tomato plant itself, tomatine is a plus. There are some indications that high levels of tomatine tend to fend off *Fusarium* wilt, a destructive fungal disease and common bane of tomatoes. Tomatine is used medicinally today in ointments for treatment of fungal skin diseases, and its presence may explain the popularity of tomato pulp as a skin salve in colonial times. Less explicably, our ancestors used tomatoes (externally) to treat glaucoma and St. Anthony's fire. Modern sources claim that an external application of tomato juice will neutralize the odor of butyl mercaptan, the nose-shriveling prime ingredient in the defense spray of skunks.

Despite all the bad press, tomatoes still managed to edge themselves inexorably into the European diet. The Italians fell for them first; by the late 1500s, tomatoes were a popular Italian dish, cooked with olive oil, salt, and pepper. (The Italians also extracted oil from the tomato seeds and manufactured tomato soap.) The English and French remained suspicious. Tomatoes were grown, but primarily as ornamentals: Henry VIII had a few as curiosities in his carnival-like garden, along with the sundials, the summerhouses, and the wooden dragons and "leberdes" (leopards) set up on green-and-white-striped poles.

The American colonists, in the matter of tomatoes, followed the negative lead of their mother country. The fruits were generally condemned by ministers and physicians. The Pilgrims considered them an abomination, on par with dancing, card-playing, and theater-going—and at least one liberal pastor, in early Massachusetts Bay, was fired by his congregation for thoughtlessly growing some in his kitchen garden.

Tomatoes were generally condemned by ministers and physicians. The Pilgrims considered them an abomination, on a par with dancing, card-playing, and theater-going—and at least one liberal pastor, in early Massachusetts Bay, was fired by his congregation for thoughtlessly growing some in his kitchen garden.

The first formal American reference to tomatoes as food plants was in Thomas Jefferson's *Garden Book* in 1781. Jefferson, an enthusiastic experimental gardener and insatiable gourmet, established his 1⅓-acre vegetable garden at Monticello in 1774, aided by an experienced Italian gardener and, optimistically, a number of vignerons from Tuscany. His wine grapes were not a success; nor was his fling at establishing an olive grove, but the vegetable garden was a masterpiece. Along with the much-maligned tomatoes, he grew carrots, radishes, lettuce, beans, peas (his favorite), lentils, celery, salsify, cucumbers, squash, eggplant, artichokes, cauliflower, spinach, endive, Spanish onions, Savoy cabbage, turnips, chicory, brussels sprouts, and sugar beets. The table at Monticello was famous. Jefferson, who styled himself an epicurean, was an enthusiastic promoter of French cuisine as well as an advocate of native American foods. He is credited with the introduction of french fries, waffles, macaroni, and Baked Alaska to this country, but Jeffersonian guests were equally likely to be fed with homestyle wild turkey or roasted squirrel. More spectacular meals ran to "Ducks, Hams, Chickens, Beef, Pigg, Tarts, Creams, Custards, gellies, fools, Trifles, floating Islands, Beer, Porter, Punch, Wine," plus, in season, peas in mint sauce, summer squash with bacon, celery with almonds, and scalloped potatoes with brown sugar. John Adams, Puritan to the core, described such repasts as "sinful feasts." There seem to be no surviving accounts of visitors' responses to the notorious tomatoes, but according to Jefferson's records, the yellow tomato crop of 1782 made an admirable preserve, similar in taste to tart apricots.

The table at Monticello was famous. Jefferson, who styled himself an epicurean, was an enthusiastic promoter of French cuisine as well as an advocate of native American foods.

Outside of Jefferson's little mountain in Virginia, acceptance of the tomato was slow. In 1798, tomatoes were introduced to Philadelphia by a vegetable-promoting refugee from Santo Domingo, but were given a tepid reception by the City of Brotherly Love. Four years later, an Italian painter brought some to Salem, Massachusetts, and was unable to persuade anybody to even taste them.

The turning point for the pro-tomato faction, according to time-honored tomato legend, occurred on September 26, 1820, on the steps of the courthouse in Salem, New Jersey, when Colonel Robert Gibbon Johnson ate, in public and without ill effect, an entire basketful of tomatoes. The Colonel, a notorious eccentric, was not a man to be trifled with. During the Revolutionary War, at the tender age of seven, he had patriotically slapped a British officer in the face, and as an adult, he habitually dressed in imitation of General Washington: a black suit with impeccable white ruffles, a tricorn hat, black gloves, and gold-topped walking stick.

Tomatoes, claimed the Colonel, had been eaten by the ancient Egyptians and Greeks but the original accounts of this beneficial diet had been lost in the mists of history. The Colonel's personal physician, a Dr. James Van Meeter, took a dim view of the proposed tomato-eating and was quoted saying, "The foolish colonel will foam and froth at the mouth and double over with appendicitis." Also threatened were aggravated high blood pressure and brain fever; barring immediate effects, it was feared that the tomato skins would stick to the lining of the stomach and eventually cause cancer. (Tomatoes were generally held to induce cancer until nearly the end of the nineteenth century.) Two thousand people assembled to watch Colonel Johnson suffer these awful fates, to the accompaniment of a local fireman's band, playing dirges. The Colonel, undaunted, ate and stalked away, to live in undisputed health to the ripe old age of seventy-nine.

Two thousand people assembled to watch Colonel Johnson suffer these awful fates, to the accompaniment of a local fireman's band, playing dirges. The Colonel, undaunted, ate and stalked away, to live in undisputed health to the ripe old age of seventy-nine.

The tomato tide was also turning on the Continent, under the prevailing influence, at least one story holds, of the French Revolution. In 1783, when Parisian citizens were wearing red caps to show that they were dedicated republicans, a patriotic chef decided to take matters a step further by serving meals solely composed of republican red food. Foremost among the republican dishes was a concoction of stewed red tomatoes, a culinary as well as a political success, which set Paris on the road to tomato cookery. An alternative story states that the tomato was introduced to French cuisine at two o'clock in the afternoon of June 14, 1800, when Napoleon's chef prepared a dish of chicken, crayfish, eggs, tomatoes, and garlic to celebrate his master's victory over the Italians at the Battle of Marengo—a recipe to be known henceforth as Chicken Marengo.

In any case, by the early nineteenth century the tomato, on both sides of the Atlantic, was coming into its own. The Landreth Seed Company in Philadelphia (possessor, according to its unbiased founder, of the "finest vegetable market in the Union") is credited with the introduction of the tomato to garden cultivation in 1820. By 1825, the Boston seed catalog of John B. Russell was offering the "Tomato or Love-Apple," along with fourteen kinds of beans, four kinds of pumpkins, and five kinds of squash (including "Akorn"). By the 1830s, Boston's Hovey & Co. carried two kinds of tomatoes ("small and large") and five years later had upped the count to four. Tomatoes were being grown in the gardens of Maine, announced the editor of the *Maine Farmer* in 1835, and were found "a useful article of diet"—probably, for Down East conservatives, gushing praise.

From the garden, the tomato naturally worked its way into the kitchen and the medicine cabinet. Physicians, both amateur and professional, touted the tomato as a remedy for indigestion, diarrhea, and liver disease, and as a cholera preventive. Popular nostrums included such cure-alls as "Dr. Miles's Compound Extract of Tomato" and "Dr. Phelps Compound Tomato Pills." Eliza Leslie's *Directions for Cookery* (Philadelphia, 1828) included several recipes for tomatoes, both alone and in company, plus one for "tomata catchup," said to make a dandy meat sauce.

Ketchup these days is a name synonymous with tomato, but the original sauce was quite literally a different kettle of fish. The modern word derives from the Malayan *ketchap,* a spicy pickled fish sauce, introduced to Europe by Dutch East India Company traders. The predominant ketchup of the eighteenth century was made with walnuts, though there were any number of popular variations. A recipe of 1760 describes a "shrimp catsup" made from shrimps stewed in gooseberry vinegar, and the versatile Miss Leslie, along with her tomato and standard walnut ketchups, included recipes for those made with lemons, mushrooms, oysters, anchovies, and lobsters. There was also a "sea catchup" designed to keep during long sea voyages.

The name that made tomato ketchup famous in the United States, however, was that of Pennsylvanian Henry J. Heinz, whose product went on the market in 1876. By 1896, Heinz had expanded his line of pickles, sauces, and relishes to the famous 57 varieties (actually something over sixty, but Heinz liked the ring of the number fifty-seven) and had erected New York City's first electric sign to that effect on the corner of Broadway and Fifth Avenue. The sign, six stories high, read "Heinz 57 Good Things for the Table" in twelve hun-

TOMATO KETCHUP

From Mrs. Samuel Witherhorne, Sugar House Book*, 1801*

◆ Get them quite ripe on a dry day, squeeze them with your hands till reduced to a pulp, then put half a pound of fine salt to one hundred tomatoes, and boil them for two hours. Stir them to prevent burning. While hot press them through a fine sieve, with a silver spoon till nought but the skin remains, then add a little mace, 3 nutmegs, allspice, cloves, cinnamon, ginger and pepper to taste. Boil over a slow fire till quite thick, stir all the time. Bottle when cold. One hundred tomatoes will make four or five bottles and keep good for two or three years.

dred electric lights, illuminated at a cost of $90 nightly.

The tomato, outside of the sauce bottle, was doing well for itself by the middle of the nineteenth century. In fact, its popularity was such that an English tourist of the 1850s, stuffed with tomatoes at every turn, commented acidly, "Its very name I now perfectly dread—so constantly, so regularly, does it come up every day, prepared in every imaginable way." By the 1870s, these dreaded tomatoes were available year-round: fresh tomatoes for city-dwellers were grown along with cucumbers and rhubarb in coal-heated greenhouses springing up in the suburban market gardens of Boston, Philadelphia, and New York. Canned tomatoes, for rural residents, became generally available in the same decade. The first commercial brand, a variety of large beefsteak tomatoes packed one to a can, was marketed by the firm of Joseph Campbell and Abraham Anderson of Camden, New Jersey. (The label on the can showed two men straining hopelessly to move a gargantuan tomato.)

Even more successfully, Campbell brought out his famous canned Tomato Soup in 1897, shortly after chemist John Dorrance, at a weekly salary of $7.59, worked out the formula for condensing it. Condensed soup was a vast taste improvement over the reconstitutable soups of the past, which were primarily available in the form of dried bricks.

Even more successfully, Campbell brought out his famous canned Tomato Soup in 1897, shortly after chemist John Dorrance, at a weekly salary of $7.59, worked out the formula for condensing it. Condensed soup was a vast taste improvement over the reconstitutable soups of the past, which were primarily available in the form of dried bricks, somewhat resembling bouillon cubes. Captain Cook took cases of such "portable soup" along on his voyage around the world in 1772—a sample slab, described as a grayish rectangular cake, was obtained from the Royal United Services Institution and analyzed in 1938. It hadn't changed much, the examining scientists stated, over the past 160 years. In 1962, Andy Warhol, who never had to subsist on portable soup, sold a painting of a Campbell's Tomato Soup can for $60,000.

The garden tomato, by the mid-1800s, had become an item of considerable horticultural enthusiasm. Tomato-breeding programs aimed at improving the garden fruit were begun during this period. These were based for the most part on breeder patience. Careful and constant observation led to selection of interesting variants, which had resulted from chance mutation, spontaneous outcrossing, or recombination of preexisting genetic traits. In a garden left to itself, such events are rare, but the pace picked up rapidly with the intervention of man, armed, early on, with a pollen-laden paintbrush. The first major breeding success of this sort was probably the Trophy tomato, developed in the 1850s by a Dr. Hand of Baltimore County, Maryland, by crossing a small, smooth-skinned ornamental tomato with a larger lobed garden variety. The product, chunky, smooth-skinned, and solid, dominated tomato patches for the next several decades and figured in the parentage of hundreds of tomato varieties grown today. Similarly successful was the Paragon tomato, offspring of a serendipitous smooth-skinned mutant fruit discovered and developed by Alexander Livingston, an Ohio farm boy. It appeared on the market in 1870. Seed catalogs of the 1880s routinely carried upwards of a dozen tomato varieties, with the number increasing annually. W. Atlee Burpee's 1888 *Farm Annual,* for example, offered fifteen large-fruited and seven small-fruited tomato cultivars, including Paragon and Trophy, Perfection, Cardinal, Mayflower, and Golden Queen. Also available was the Faultless Early Tomato, which, Burpee stated with admirable objectivity, was far from faultless, the fruits being disappointingly rough. By 1901, Burpee, who was certainly in a position to know, announced, "The tomato now rivals all other vegetables and fruits in popularity, having reached a use beyond that of the potato and apple combined."

All this past and present popularity is a pity in some ways, since the tomato, nutritionally, is no great shakes. Though plugged as the "oranges of the vegetable garden" for their reputedly high vitamin C content, tomatoes are actually unimpressive. A tomato provides only about one-third as much vitamin C as a green pepper or a cantaloupe half, one-fourth as much vitamin C as a cup of orange juice, and one-fifth as much as a stalk of broccoli. In a comparative study of vegetables and fruits performed by M. A. Stevens at the University of California at Davis, tomatoes were found to rank sixteenth in overall concentration of ten selected vitamins and minerals, considerably behind such traditional vegetable anathemas as spinach (#2) and lima beans (#4). (Top of the list was broccoli.) As specific sources of vitamins C and A, tomatoes ranked thirteenth and sixteenth, respectively.

The tomato makes up for its deficiencies in nutritional quality, however, by the quantities in which we consume it. Stevens found that tomatoes ranked number one in terms of their contribution of nutrients to the American diet, simply because we eat a lot of them. Tomatoes, to their credit, are among the foods that weight-watchers can eat a lot of with a clear conscience: they contain 93.5 percent water (only cucumbers and a few leafy vegetables contain more) and log in at a piddling four calories per ounce. That means, for the caloric price of a chocolate ice cream cone, you can wolf down about three hundred cherry tomatoes. Nutritionally, it's also possible to stack the deck by using some care in selecting your tomatoes. Homegrown types, for example, ripened all the way on the vine, have about one-third more vitamin C than the artificially ripened supermarket varieties—and, better than that, there are some tomato cultivars bred for unusually high concentrations of vitamin C. One of these, Doublerich, an early red medium-sized tomato, contains twice the C of most. Most of the vitamin C, in all tomatoes, is concentrated in the jelly-like material in the middle, surrounding the seeds.

The tomato makes up for its deficiencies in nutritional quality, however, by the quantities in which we consume it. Stevens found that tomatoes ranked number one in terms of contribution of nutrients to the American diet, simply because we eat a lot of them.

Artificial tomato ripening, connoisseurs and common folk agree, is a sin against more tomato attributes than just vitamin C. Few foods are as delicious as a ripe, fresh-picked garden tomato; conversely, few are as unspeakably dreadful as the wholly, anemic substitutes found on supermarket shelves in mid-December. These abominations are produced by picking tomatoes in the green stage, then exposing them to ethylene gas, a process delicately known in the tomato industry as "de-greening." Ethylene, primarily a breakdown product of the amino acid methionine, is present normally in fruits and acts much like a hormone, triggering the many activities that lead to natural ripening. It can also lead to unnatural ripening. The ancient Chinese were aware that fruit would ripen more rapidly if placed in a closed chamber with burning incense. In this country, considerably later, orange growers noticed that green oranges placed in rooms with oil heaters rapidly turned into orange oranges. The reason for this color change, at first assumed to be simply heat, was soon discovered to be the presence of ethylene gas, an incomplete oil combustion product. Tomatoes are much more resistant to ethylene than are oranges, but will, if determinedly treated, eventually turn reddish. The unfortunate result is solid proof that, like trees, only God can make a tomato.

Attempts to bypass the Creator through artificial hybridization multiplied astronomically in the 1920s and 30s, and countless tomato cultivars have come (and gone) since. The most spectacular success of the hybridizers, in the view of modern commercial tomato growers, has been the development of a tomato suitable for mechanical harvesting. Much of the unprincipled effort in the making of the supermarket tomato went on at the University of California at Davis in the 1950s, and the result, a fruit designed for uniform ripening, thick skin, firm flesh (you can play catch with it without hurting it), and the ability to respond well in the ethylene chamber, has won high marks from California's multimillion-dollar tomato industry. A further improvement has been the creation of the "square" tomato, a chunky, compact fruit better suited to packing and shipping than to eating. If you're looking for a vegetable with potential as a food-fight projectile, this is your tomato. Talk now is of developing a bluntly elongated sausage-shaped tomato the approximate diameter of a hamburger, suitable for slicing.

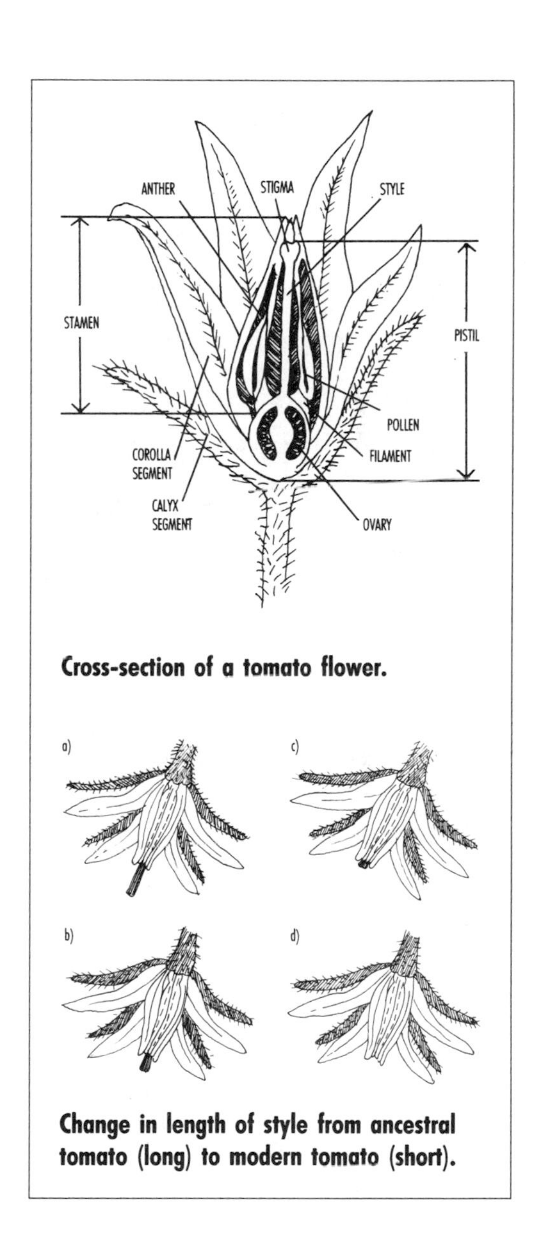

Cross-section of a tomato flower.

Change in length of style from ancestral tomato (long) to modern tomato (short).

The making of a tomato, whether by human hand or by Mother Nature, begins with the fertilization of the female ovule by the male pollen. In the modern tomato, this is an incestuous process known as self-fertilization, a practice adopted by the plant soon after its introduction to Europe. Tomato forebears, the wild South American ancestral tomato and the Central American cultivated varieties, were routinely crosspollinated, by either wind or pollen-toting insects. To facilitate this, the original tomato blossom possessed an extended (exserted) stigma—the smokestack-like tip of the pistil, which in turn is the female organ containing the ovary. The exserted stigma sticks up well beyond the anther cone, where the pollen is produced. It is thus more likely to contact pollen grains from eager neighbors than those from the home source, which, to hit the target, would have to fall up. Pollen grains land on the sticky surface of the stigma and germinate, extending long tubes downward through the style to the ovary. Down these tubes travel a pair of fertilization-bent male nuclei. One fuses with the female "egg" to form the seed embryo; the other fuses with an adjacent cell to form the endosperm, future food for the developing infant plant.

All this proceeded as expected in the familiar insect-populated climate of Mexico; when transported to Europe, however, the abrupt dearth of pollinating insects left the tomato in a difficult position. Eventually the European tomato solved its sexual problems by reducing style length, retracting until the stigma in modern plants sits well down within the anther tube. This shift in position allows self-fertilization—in fact, makes it a virtual necessity—and in turn increases the yield of fruits per tomato plant.

All this proceeded as expected in the familiar insect-populated climate of Mexico; however, when transported to Europe, the abrupt dearth of pollinating insects left the tomato in a difficult position.

The tomato is an unusual plant in that cell division in the future fruit is over almost at the moment of fertilization. A tiny, but fully formed, infant tomato can be seen at the base of the flower as soon as it opens. Further development is largely a matter of cell growth—usually the tomato reaches full size in twenty to thirty days, about half the length of the total ripening period. During this time, the growing fruit accumulates quantities of water, minerals, and starch, and new cell-wall material is laid down. Cell-wall material consists of cellulose—the primary stiffening component in plant cells, digestible only with difficulty by cows and termites, and not at all by people—embedded in an equally tough layer of insoluble cement-like pectin. Cellulose and insoluble pectin are the main contributors to the green tomato's crunchy texture.

Once the tomato reaches peak size, it passes through a one- to two-week maturation period, during which starch storage continues and a number of developmental changes pave the way for the nitty-gritty of ripening. Mature peak size varies considerably from cultivar to cultivar, ranging from huge to tiny. On the huge end are the Bragger, the Big Boy, the Oxheart, the Mammoth Wonder, the Watermelon Beefsteak, which puts out pink two-pounders, and the Giant Belgium, said to yield spectacular fruits of up to five pounds. Also known as the sweet wine tomato, this last is said to be the best bet for whipping up a batch of tomato wine. It comes not from Belgium, but from Ohio. According to Robert Hendrickson in *The Great American Tomato Book,* the world record for humongous tomatoes was held for years by a grower from Calaveras County, California, who in 1893 turned out a 4½ pound, 8-inch-diameter colossus. In 1976, according to the *Guinness Book of World Records* (established in 1956 by the owners of the Guinness Brewery in hopes of "providing a means for the peaceful settlement of arguments"), a new record-breaker appeared on the scene, a 6½-pound tomato from the garden of Clarence Dailey of Monona, Wisconsin. At the opposite end of the scale are the cherry tomatoes, with many names reminiscent of miniature poodles: Toy Boy, Tiny Tim, Sweetie, Small Fry. Fruits may be as small as half an ounce and ¾ inch in diameter, a mere drop in the bucket compared to Mr. Dailey's prize specimen. Still, you can eat them all in one bite without the juice dripping down your chin.

At whatever size, as growth and maturation are completed, ethylene production goes up and in tomatoes there is an abrupt rise in respiration, which, scientists tell us, signals the beginning of the end in the life of a fruit. This respiratory upsurge, termed a climacteric rise, is displayed by a number of fruits, the cantaloupe, honeydew melon, watermelon, peach, pear, plum, and apple, as well as the tomato. (Nonclimacteric fruits, among them oranges and lemons, ripen without an initial respiratory skyrocket.)

Initially the fruit lightens in a star-shaped pattern at the blossom end—the "starbreaker" stage, to those in the tomato trade. It then proceeds gradually through yellowish-pink to orange to the deep rich red ordinarily associated with the mouth-watering ripe garden tomato.

About two days after the tomato reaches the mature green stage, the most obvious of the ripening processes, the color change, commences. Initially the fruit lightens in a star-shaped pattern at the blossom end—the "starbreaker" stage, to those in the tomato trade. It then proceeds gradually through yellowish-pink to orange to the deep rich red ordinarily associated with the mouth-watering ripe garden tomato. Chemically, this color change is due to the breakdown of chlorophyll (green) and the synthesis of carotenoids (yellow and red). Tomato-red is largely due to a carotenoid pigment called lycopene, the same agent that colors the flesh of pink grapefruits and watermelons. The reddest of red tomatoes is said to be a high-lycopene variety called Trimson, developed at the University of Toronto from a wild species found in the Philippines. Other color variations in tomatoes also result from genetic alterations affecting pigment production. White tomatoes, for example, degrade chlorophyll normally but synthesize no carotenoids. Such behavior is governed by a genetic locus termed *gh* for ghost. Yellow tomatoes produce yellow carotenoids, but no red lycopene; orange tomatoes are similarly lycopene-less, but high in the carotenoid beta-carotene, which also puts the orange in carrots and sweet potatoes. There is even a variety of tomato that ripens, but does not degrade chlorophyll and so remains confusingly green. It is called, appropriately, Evergreen.

Flavor in any fruit is a complicated mix of sugars, organic acids, and many miscellaneous volatile compounds—at least 118 in the ripe tomato and over 200 in the equally ripe banana.

Along with the color change, ripening involves marked changes in texture and taste. The fruit softens due to the activity of enzymes, pectinesterase and polygalacturonase, which convert through the insoluble cell-wall protopectins to soluble form. In the unripe fruit, the insoluble pectins act to strengthen cell walls and to bind adjacent cells together; soluble pectins, on the other hand, weaken the whole structure and allow the cells to separate easily when bitten. In the absence of pectinesterase and polygalacturonase, tomatoes would have to be gnawed. There are some tomato mutants that suffer from just that: one of these, designated Neverripe (Nr), produces only miniscule amounts of the required enzymes and softens extremely slowly. Understandably, scientists are more interested in it than are gardeners.

Finally, and most important, ripening involves the development of true tomato taste, the quality we all fantasize about, but don't get, from the winterbound A&P. Flavor in any fruit is a complicated mix of sugars, organic acids, and many miscellaneous volatile compounds—at least 118 in the ripe tomato and over 200 in the equally ripe banana. In the green tomato, most of the sugar is stashed in the storage form of (unsweet) starch. As ripening progresses, the enzyme alpha-amylase—found fulfilling the same function in human saliva—rapidly hydrolyzes this starch, converting it to (sweet) glucose and fructose. Simultaneously, the concentration of (sour) organic acid drops off, and the result, a nice balance of sugar, acid, and volatiles, is what makes up the perfect tomato. Acidity, in the average perfect tomato, generally ranges around pH 4.0 to 4.5 (pH 7.0 = neutrality), in the same ballpark as red cabbages, onions, and pears. Lemon juice, in contrast, logs in at a puckering pH of 2.3, and vinegar at 2.5.

Though not as sensitive as the fabled five-minute sweet corn, tomatoes also should be eaten fairly soon after picking, since sugar decreases in storage. The decrease is probably due to consumption during respiration. Fresh fruits and vegetables, plant scientists remind us testily, are living things. As tomatoes sit lumpishly on the kitchen shelf, they *breathe*. They should not be kept in plastic bags.

Though not as sensitive as the fabled five-minute sweet corn, tomatoes also should be eaten fairly soon after picking, since sugar decreases in storage.

The tomato, botanically, is a fruit—that is, an organ that develops from the ovary of the flower and encloses the developing seeds. The fruit of the tomato, like the avocado and papaya, is more specifically a berry, composed of seeds surrounded by parenchymatous cells. Unlike most, however, tomato fruits have been given a legal as well as a botanical definition. The legal tomato appeared in the late nineteenth century, when, in 1886, an importer named John Nix landed a load of West Indian tomatoes in New York. The presiding customs agent promptly levied a 10 percent duty on the load, according to the regulations set down by the Tariff Act of 1883. Nix, who knew his botany, protested that the tariff applied only to vegetables; tomatoes, as fruits, should be exempt. The fruit-vegetable controversy was bandied about for some years and finally, in 1893, reached the Supreme Court. There, Justice Horace Gray ruled in favor of vegetable: "Botanically speaking, tomatoes are the fruit of the vine, just as are cucumbers, squashes, beans and peas. But in the common language of the people . . . all these vegetables . . . are usually served at dinner, in, with, or after the soup, fish, or meat, which constitute the principal part of the repast, and not, like fruits, generally as dessert." Nix paid up.

The fruit, despite much gardener opinion to the contrary, is not an end in itself, but a tomato plant's clever way of making more tomato plants—a fancy mechanism for seed dispersal. Within each tomato fruit (or vegetable) are 250–300 tiny seeds, which weigh in at about five thousand to the ounce. These seeds develop most rapidly during the second half of the tomato maturation period, the one- to two-week pause between the attainment of full growth and the onset of ripening. During this time, the seed embryo reaches full size and the seed coat develops and hardens. Animal-assisted seed dispersal, of the sort aspired to by the tomato, is known as endozoochory, "seeds inside animals," or what one researcher terms the "Jonah syndrome." Here, the seeds are covered by an appealing coat of fleshy food and, at some point during the eating process, are spat out, spilled, or voided (dispersed) by the cooperative animal. For this dispersal to be effective, the seeds must escape wholesale digestion and destruction by overenthusiastic eaters, a problem the tomato gets around by producing immense numbers of small seeds. Some inevitably spill during feeding and go on to reproduce the species, and those that are actually swallowed possess coats tough enough to resist the fatal activity of the digestive tract enzymes. The seeds of at least one species, the wild tomato of the Galápagos Islands, have seed coats so indestructible that they are unable to germinate unless first partially digested in the gut of the Galápagos tortoise.

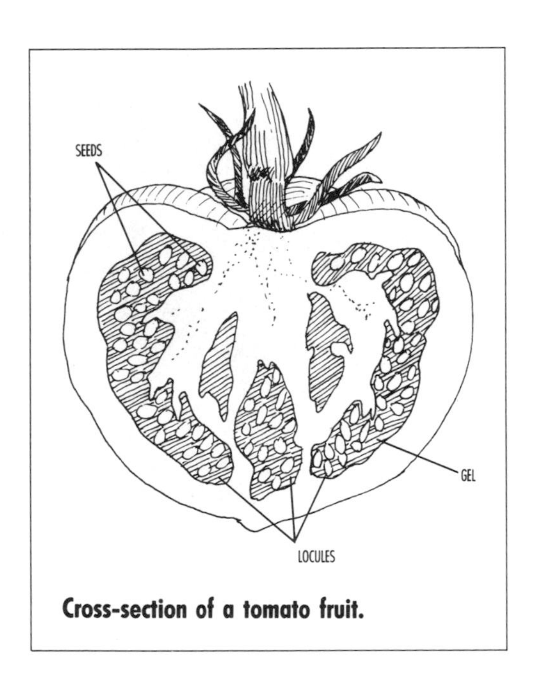

Cross-section of a tomato fruit.

A whole literature on tomato-growing has sprung up in recent years, with an abundance of helpful hints for both beginning and experienced gardeners. Many of these were already helpfully at hand in the nineteenth century. Modern organic gardeners recommend fertilizers of bone meal, rich in phosphorus; animal manure, rich in nitrogen; and wood ashes, a source of potassium—all three liberally used in the last century. Nineteenth-century coastal tomato-growers further beefed up their plants with a compost of seaweed, observed by Henry David Thoreau around the tomato plants of Marshfield, Massachusetts. Horse, chicken, rabbit, and sheep manures are said to be the best for tomato cultivation, but, if you can lay a shovel on it, elephant manure, according to the Ringling Brothers and Barnum & Bailey Circus, grows "the biggest and best tomatoes in the world." Tiger manure, though not specifically recommended for tomatoes, is said to ward off marauding gophers and rabbits if spread around the edge of the garden.

Some gardeners claim that music motivates their tomatoes, a statement hesitatingly supported by scientists, who suggest that sonic stimulation may indeed be a valid horticultural technique. Early tomato growers, unless they whistled while they worked, seem not to have opted for this technique, but garden music certainly has a substantial history. Nero fiddled in his garden, medieval estate owners routinely kept groups of musicians to play in theirs, and Samuel Pepys used to sing in his (to the dismay of the neighbors). Perhaps it all helped the plants. W. C. Fields, hater of children and dogs, used to cuss at his tomato plants, which is said to be bad, though bad temper also has its horticultural place. Basil, since Roman times, was thought to grow better if planted with curses and hatred—which of course may be true, though mine, usually planted in a pleasant mood, seems to do well.

Tomato cages were already in use in nineteenth-century gardens, constructed by driving three stakes into the ground around the plant, then circling the stakes with barrel hoops. Tomato cages are still about the best culture method going—one study found single plants in cages yielded up to sixty-two pounds of fruit. Even better is a structure exotically called the Japanese ring, in which the plants are grown outside a wire cage. Using these, yields have been recorded of up to one hundred pounds per plant. There is some evidence today that the prime crops in the tomato cages may be due to the growth-stimulatory effect of the electromagnetic field set up by the wire cage frames. Even this is not a completely new idea: nineteenth-century gardeners thought "magnetism" a plus for developing plants and invested in "electrical strips" of copper and zinc, which were arched like croquet hoops over sprouting vegetables. Barring cages, it's best to simply let the plants sprawl. Though not as productive as their caged buddies, the sprawling vines still produce about twice as many tomatoes as staked plants, though they look

Horse, chicken, rabbit, and sheep manures are said to be the best for tomato cultivation, but, if you can lay a shovel on it, elephant manure, according to the Ringling Brothers and Barnum & Bailey Circus, grows "the biggest and best tomatoes in the world."

messier and you tend to step on them while harvesting.

With proper plant selection, it may not be necessary to make the stake-or-sprawl decision at all. Bushy determinate plant types, for example, grow no more than three feet tall and require no staking. Determinate plants grow in a compact, tidily circular fashion, a habit governed by a specific gene designated *sp* (self-pruning). This characteristic appeared out of the blue as a spontaneous mutation in Florida in 1914, and has been a boon to tomato breeders ever since. The branches of *sp* plants all terminate at about the same distance from the main stem, ending in flower buds. In contrast, the indeterminate tomato vine grows indefinitely, continually sprouting new leaves at the branch tip. If not staked, the indeterminate plant will continue to crawl across the ground in all directions, growth only grinding to a halt with the advent of unseasonably cold weather.

The Siberia tomato sets fruit at 38 degrees F and produces ripe fruit in forty-eight days flat after flowering.

For the tomato, basically a warm-weather plant, cold weather usually isn't all that cold. Most tomatoes prefer a soil temperature of 70–80 degrees F and a daytime air temperature of 65–80 degrees F—and prolonged exposure to temperatures under 50 degrees F is enough to do most tomatoes in. During the growth season, the result of such unfortunate chilliness is the failure of the tomato flowers to set fruit. Instead, an abscission layer forms, of the sort that separates deciduous leaves from their branches in autumn, and the flowers fall to the ground. Generally this happens because the flowers have not been successfully fertilized. In many cultivars, at temperatures below 55 degrees F, pollen germination and subsequent tube growth are so slow that the male nuclei cannot manage to make contact with the female before the blossoms detach. Such cold sensitivity is doubtless why the Pilgrims didn't find tomatoes flourishing in the fields of the New England Indians along with the corn, squashes, and beans. However, the tomato being a versatile plant, there are cold-hardy exceptions—noteworthy among them the Siberia tomato, which arrived in Canada in the 1970s by way of a Russian visitor from Siberia. The Siberia tomato sets fruit at 38 degrees F and produces ripe fruit in forty-eight days flat after flowering. It has been successfully grown in the Yukon and Alaska, has won first prize in a horticultural contest in Newfoundland, hardly tomato country, and has been grown outside in the dead of winter in southern Florida and California.

Today there are well over five hundred tomato cultivars on the market, adapted to any number of climates, with new varieties introduced yearly. In 1885, the family Vilmorin-Andrieux, the famous French seedsmen, listed in their magnum opus, *The Vegetable Garden,* five varieties of ribbed tomatoes; four of round tomatoes, including the colorful Apple-Shaped Red, Apple-Shaped Rose, and Apple-Shaped-Purple; two pear-shaped or fig tomatoes; one plum tomato; one cherry tomato (yellow); the curiously shaped King Humbert, flat on four sides and in cross-section nearly square; the Turk's-cap tomato, which bore a topknot-like protuberance in the middle much like that of the turban gourds; and the tree tomato, cultivated in the garden of the Count de Fleurieu at the Château de Laye. Sold in this country a century ago as Laye's Upright Red Tree tomato, this plant, *Cyphomandra betacea,* is a native of Peru and a relative of the garden tomato. In its native habitat, the tree tomato, a perennial with elephant-eared foot-long leaves and purple or green flowers, reaches heights of eight to twelve feet. It bears brown egg-sized and egg-shaped fruits seven months out of the year, vaguely tomato-tasting and said to smell like carnations. Nowadays it is cultivated primarily in Ecuador.

Also described by Vilmorin-Andrieux was the strawberry, or husk, tomato, a tomato relative belonging to the genus *Physalis,* which contains some eighty species. These include both ornamentals, such as the Japanese- or Chinese-lantern plant (*P. Alkekengi*), whose papery bright-orange calyces are used in dried-flower arrangements, and edibles, such as the tomatillo or jamberry (*P. ixocarpa*) used in the Mexican *salsa verde.* The husk tomato most commonly grown as an edible garden vegetable is *P. pruinosa,* which bears a canary-yellow fruit encased in a brownish husk. The fruit does not ripen until it falls to the ground, hence its nickname ground cherry. Similar are *P. peruviana*—in England called the Cape gooseberry, because it was popularly grown in South Africa near the Cape of Good Hope—and the slightly smaller *P. pubescens.* Edible husk tomatoes have been grown in this country since the 1850s and traditionally used as dessert fruits, in pies or preserves or dried in sugar and eaten like raisins. In Hawaii, *P. pruinosa,* the poha berry, is used to make yellow poha jam.

Not mentioned by Vilmorin-Andrieux, and not cultivated in this country to my infinite regret, is a tomato relative native to the Fiji Islands known scientifically as *Solanum anthropophagorum,* the cannibal's tomato. Described in 1840 by Charles Wilkes, on an official expedition of exploration for the U.S. Navy, this tomato, a small yellow egg-shaped fruit, was routinely eaten by the Islanders as a stomach-settling chaser after a main meal of missionary.

Researchers have also identified a potentially valuable salt-tolerant tomato species on the Galápagos Islands—it grows on the coast, five yards above the high-tide line, and can survive in seawater—and a drought-resistant variety in western Peru, which essentially survives on what water it gets from fog.

Our own garden tomato, *L. esculentum,* is one of nine species belonging to the genus *Lycopersicon,* and the only one grown to any extent for human consumption. (*L. pimpinellifolium,* the currant tomato, is also grown, but rarely.) Still, wild tomatoes flourish the world over, variously described by plant taxonomists as weak, weedy, clambering, coarse, and distasteful. Unaesthetic though they are, these wild varieties constitute a vast and valuable genetic resource. Dr. Charles Rick of the University of California at Davis says that resistance to at least twenty-seven serious tomato diseases has been discovered in the wild, and that many of these resistant traits have been bred into previously susceptible domestic tomatoes. Researchers have also identified a potentially valuable salt-tolerant tomato species on the Galápagos Islands—it grows on the coast, five yards above the high-tide line, and can survive in seawater—and a drought-resistant variety in western Peru, which essentially survives on what water it gets from fog. A continuing worry among plant geneticists is that such species and traits, many of them already endangered, will vanish before their best features can be incorporated into the domestic gene pool.

Tomato improvement techniques have become more sophisticated in recent years, with the movement of plants out of the garden and into the laboratory. One of the latest methods of generating new tomato cultivars employs plant-tissue culture techniques to exploit a phenomenon known as somaclonal variation. Plant-tissue culture techniques have been around in greater or lesser degrees of sophistication for the last fifty years. Basically, segments of selected plant tissue—usually chunks of leaves or stems—are minced to release individual cells, which are then sterilely grown up in the laboratory in specially formulated solutions of nutrients and hormones. Each of these cells, by manipulations of the culture medium, can be regenerated into a whole new plant, identical to the original parent. In practice, however, often an annoyingly far cry from theory, the regenerated plants frequently show a marked array of differences. This somaclonal variation has been observed in over thirty plant species, among them potato, celery, lettuce, and carrot, as well as the tomato.

Identical-clone-minded scientists viewed somaclonal variation as a howling nuisance until 1981, when Australian researchers at the Commonwealth Scientific and Industrial Research Organization (CSIRO) in Canberra suggested that it might serve as a superb source of new plant traits. In this country, the technique has been used, primarily by DNA Plant Technology (DNAP) of Cinnaminson, New Jersey, to identify and isolate a new high-solid tomato suitable for the processing industry. The research has been funded by the Campbell's Soup Company, which alone uses eight hundred thousand tons of processing tomatoes annually. The current processing tomato has a solids content of only 10 percent; optimistic geneticists are hoping for a fifty percent increase. With such tools in hand, the industry also hopes for tastier tomatoes—perhaps even, one of these days, a winter tomato that tastes less like orange wool and more like the real McCoy. It's a happy thought.

And finally: the correct pronunciation, linguists believe, is to-*mah*-to, from the sixteenth-century Spanish *tomate*. The word picked up its *o* in eighteenth-century England, where the insular English believed that all Spanish words ended that way, but retained its short *a*. We, however, will continue to plant and eat to-*may*-toes. After all these years of phonetic error, it's just too blasted late to call the whole thing off.

PEPPERS

Montezuma, in best Aztec tradition, drank his cocoa cold, unsweetened, and laced with vanilla and ground hot peppers. The Spaniards, who short-sightedly described the Aztec *chocolatl* as a "food fit only for pigs," were more enthusiastic about the fiery red spice, seeing in it a substitute for the outrageously expensive black pepper, *Piper nigrum*. Black pepper, imported since ancient times from India, had served Europe for some two thousand years not only as a taste-tingling spice but also as a legitimate medium of exchange. Rome paid off the marauding Visigoths and Huns with pepper (three thousand pounds of it), and pepper featured heavily in rafts of medieval wills and dowries. Rents were settled in peppercorns, and in feudal France, if a serf could come up with a pound of black pepper, he could buy his freedom. In the 1500s, black pepper was literally worth its weight in gold, and the quest for it figured largely in the New World voyages of discovery. Spanish thinking on this point was so wishful that the fruit of the American *Capsicum* was christened *pimiento,* "pepper."

The new spice, unlike most of the New World plant products, was an instant hit. (Even chocolate, irresistible to the modern palate, was received with suspicion, and only took off in popularity after some unknown culinary innovator had the bright idea of beefing it up with sugar.) Capsicums were growing in Spanish monastery gardens by the end of the fifteenth century, and by the first half of the sixteenth had spread to Italy, France, and Germany, where they were stubbornly referred to as Indian or Calicut peppers. Conversely, India, which obtained the plants at about the same time, called them Pernambuco peppers, after a region of Brazil. By the 1560s, capsicums reached the Balkans where they were called *peperke* or *paparka*—and the Hungarians, by a short linguistic jump, had acquired their famed *paprika* by 1569. The English had capsicums, under the name of Ginnie Pepper in the 1540s, and as promptly as possible, in a sort of vicious vegetable cycle, re-introduced them to America via the early colonists.

Rather than Calicut or Ginnie, botanists now state that the capsicums, like tomatoes, originated in South America, where today there exist some twenty wild species. A more specific theory holds that all twenty evolved from a single ancestor originally located in the mountains of central Bolivia. It seems likely that peppers were domesticated simultaneously and independently in several different South and Central American locations, with the regional domesticated breed derived from the most prevalent wild species. Archaeologists have dredged up pepper seeds from Mexican cave dwellings dating to 7000 B.C., and the plants seem to have been under cultivation by some time between 5200 and 3400 B.C.

By the 1550s, botanist-physician Rembert Dodoens announced that the new peppers were strong enough to kill dogs, and in 1772 the botanically minded Dominican priest, Francisco Ximenez, wrote of a Cuban pepper so inflammatory that a single pod could render "a bull unable to eat."

Along with corn, beans, and squash, the capsicums were among the first plants cultivated in the agriculturally revolutionized Americas. There are five domesticated species on the market today, and none, when it comes right down to cases, is all that different from those selected and tended by the pre-Columbian Indians. Most common commercially these days are breeds of *Capsicum annuum,* a versatile crew, including the sweet bell peppers, the red paprika peppers, the pimiento peppers (used primarily to stuff olives), and an array of hot peppers, among them the familiar jalapeño and the "insanely hot" bird pepper. *C. frutescens* is known to most of us as the prime ingredient of Tabasco sauce; *C. chinense,* a native not of China, but of the West Indies, includes the bright yellow-orange habañero pepper, the hottest readily available. *C. baccatum* and *C. pubescens* were cultivated by the Incas of Peru, and were among the gifts with which the Inca king attempted to buy off Pizarro. *C. puebescens,* of which the Rocoto pepper variety is described as "murderous" and, by latter-day Peruvians, as "hot enough to kill a gringo," didn't do the trick; Pizarro accepted his peppers, macaws, llamas, gold, and silver, and went on to destroy the Inca civilization.

While cultivated pepper breeds differ greatly in size, shape, color, and pungency, they do show some classic common differences from their wild forebears. Wild peppers are generally tiny, red, and hot, and usually have erect fruits: that is, the pepper is attached to the parent plant in an upright position, sticking up out of the foliage, the better to attract the attention of seed-dispersing birds. In the hands of early pepper growers, who were anti-bird, plants were selected for pendent fruits, dangling downward and sneakily hidden in the leaves. Similarly, early growers selected for non-deciduous plants, in which the peppers stay firmly attached to the calyx and stem (peduncle). This is another bird-thwarting device; the wild deciduous capsicums are easily separated from their calyces, easy pickings for bird beaks, and a plus for seed dispersal, which in the boonies is largely performed by pepper-munching birds.

The various capsaicin compounds have different effects on the hapless human mouth: three give what is described as "rapid bite sensations" in the back of the palate and throat; two give a low-intensive slow burn on the tongue and mid-palate.

By the time of the Spanish invasion, the Mexican Indians had named their domesticated pepper *ají* (pronounced ah-hee), a name quickly adopted and still used in Spain today. Our word *capsicum* arrived on the scene in 1700 under the auspices of Joseph Pitton de Tournefort, early plant taxonomist and plant-hunter for the spectacular gardens of Louis XIV. The name is thought to come from either the Latin *capsa* ("box") for the, at least in some cases, box-like shape of the fruit, or the Greek *kapto* ("to bite") for the pepper's tongue-searing pungency. The latter certainly is what most often springs to mind in conjunction with pepper. The earliest known description of the American capsicum, written in 1493 by Peter Martyr, barely after the plant touched down on European soil, mentions modestly that the fruits were "pepper more pungent than that from Caucasus"; by the 1550s, botanist-physician Rembert Dodoens announced that the new peppers were strong enough to kill dogs, and in 1772 the botanically minded Dominican priest, Francisco Ximénez, wrote of a Cuban pepper so inflammatory that a single pod could render "a bull unable to eat."

These effects are due to a family of flavorless,

odorless, but explosively obvious chemical compounds known as capsaicins. At least five of these, officially known as vanillyl amides, have been identified to date—the first crystallized and named by L. T. Thresh, an Englishman working in India, where the dominant seasoning, curry, teams with capsaicin molecules. The various capsaicin compounds have different effects on the hapless human mouth: three give what is described as "rapid bite sensations" in the back of the palate and throat; two give a low-intensive slow burn on the tongue and mid-palate. Different combinations of these produce the different hotness characteristics of individual pepper strains. Recently, researchers at Maryland's McCormick & Co., largest spice and seasoning producer in the world and master-maker of hot pepper sauce, used high-performance liquid chromatography (HPLC) to measure levels of capsaicin compounds in certain pepper types. Using the results, they were able to successfully predict the hotness "bouquet" of various capsaicin mixes. From this, it may be a short step to the computerized hot sauce: mathematically precise proportions of capsaicins suited to the mild, medium, and hot pepper consumer.

Pepper pungency is expressed in Scoville Heat Units, a subjective rating in which the sweet bell pepper possesses 0 heat units, the feisty jalapeño usually ranges from 2,500 to 4,000, and the unspeakable Tabasco kicks in at 60,000 to 80,000.

Pepper rating, though perhaps soon to become the province of mechanized chemical analyses, has traditionally been a function of the professional human tongue. In these "organoleptic" tests, formally begun in 1912, a pepper sample is steeped overnight in ethanol, then the extract diluted in a sucrose solution to yield a "cordial." Samples of these cordials are tested by trained tasters, who determine the weakest dilution at which the hotness sensation is detectable. Pepper pungency is expressed in Scoville Heat Units, a subjective rating in which the sweet bell pepper possesses 0 heat units, the feisty jalapeño usually ranges from 2,500 to 4,000, and the unspeakable Tabasco kicks in at 60,000 to 80,000. Capsaicin content also is affected by climatic conditions, geographic location, and age of the fruit: in general, warm-weather peppers are higher in capsaicin than their cool-weather relatives, which explains why those crops raised in this country are as a rule calmer than those raised south of the border. The weather factor can be felt on even a short-term basis. According to Jean Andrews, Texas pepper expert, a summer heat wave will spice up the capsicums of every garden in its path. Particularly effective are sweltering nights: a high night temperature appears closely correlated to capsaicin level. Hotness also increases with age of the fruit. Infant peppers are universally harmless; in most, pungency only begins to develop around four weeks of age, then increases steadily with advancing maturity.

The capsaicin-induced burn experienced by pepper-eaters is sensed not by the taste buds, but by pain receptors in the mouth, and is detectable—as a sensation of warmth—at dilutions down to 1 part per million. Why human beings line up to consume a food inherently painful is a mystery, but psychologist Paul Rozin suggests that pepper-eating, like watching horror flicks, is an example of "constrained risk": the body responds with exciting warning signals, but the situation is not really dangerous. Still, the situation has its repellent properties. Capsaicin-laden pepper figures as an ingredient in the anti-dog-and-mugger aersols toted by mail carriers and joggers; ranchers smear it on their sheep to discourage wolves and coyotes; gardeners sprinkle it on their flower bulbs to ward off squirrels and, in Texas, around their vegetable gardens to fend off armadillos. It is featured in modern and historical organic insect sprays: in the mid-1800s, a Dr. Barton of Philadelphia boasted of successfully defeating cucumber beetles with his personal mix of red pepper and tobacco, and a rose lover of the same era advised those with insect-infested bushes to "take a shovel of live coals of fire, split open a red pepper and lay on the coals, and hold so that the smoke will go through the bush." In a brief surge of creativity in 1983, New York City transit officials dusted hot pepper on subway token slots to prevent unprincipled teenagers from sucking tokens out of the turnstiles; and, in what can only be hoped was an even briefer surge in the eighteenth century, entertainment-minded Londoners sneakily poured hot pepper into snuffboxes as a practical joke. As jokes go, it must have been a corker.

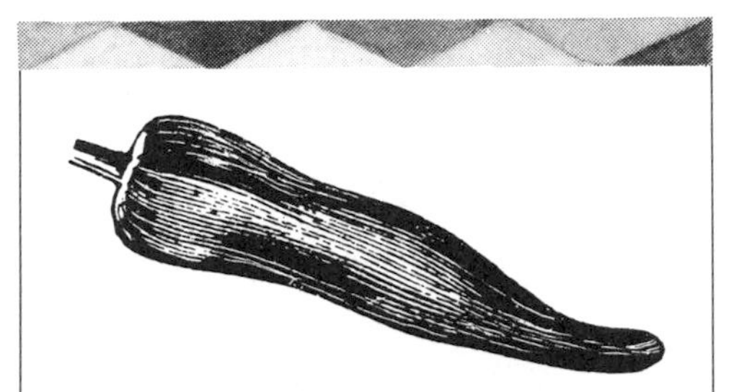

In a brief surge of creativity in 1983, New York City transit officials dusted hot pepper on subway token slots to prevent unprincipled teenagers from sucking tokens out of the turnstiles.

As well as harmlessly painful, the better peppers are also flavorful, and these days most growers and users aim for a more positive gustatory effect than a screech followed by a gulp of water. We detect flavor through the taste buds on the upper surface of the tongue, small nipple-shaped clusters of forty to sixty specialized epithelial cells primed to pick up one of the four primary tastes: the sweet, the salty, the sour, or the bitter. Adults usually possess several thousand of these taste buds—about half as many as the average pig. Both number and sensitivity of taste buds decrease after the age of forty-five, which is why the fifty-seven-year-old Queen Victoria complained that strawberries no longer tasted as delectable as they had in the days of her youth.

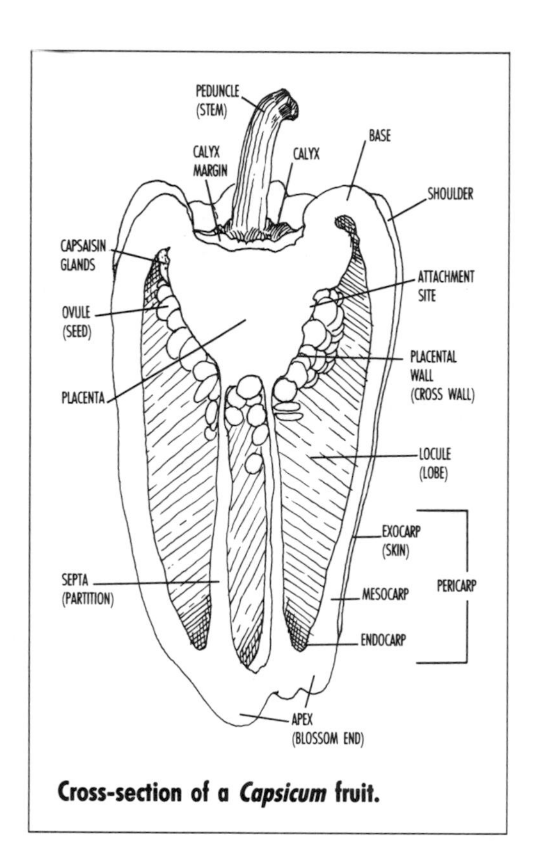

Cross-section of a *Capsicum* fruit.

Taste buds of specific types have specific locations on the tongue. Sweet-sensitive buds are located at the tip, which is why it is best to lick a lollipop or an ice-cream cone; bitter-sensitive buds are spread in a band across the back; salt- and sour-sensitive areas are arranged respectively around the front and sides. The center of the tongue is relatively taste-bud-free, which means that should you have one of those proverbial bitter pills to swallow, smack in the middle is the place to put it. Some recent evidence indicates that the capsaicin compounds inhibit perceptions of sour and bitter, which should be all to the good for the pepper flavor complex, an unknown mix of aromatic substances concentrated in the pepper's outer wall. The majority of the capsaicin compounds (89 percent), in contrast, are found in the placenta, the inner cross walls in which the seeds are embedded. Accordingly, hot peppers can be considerably toned down by scraping out the insides.

Flavor appears closely associated with the carotenoid compounds, colored molecules that turn the capsicums their gaudy reds, oranges, and yellows. The deeper the color, according to Jean Andrews, the more flavorful the pepper. The major red in red peppers comes from the carotenoid capsanthin, which constitutes some 35 percent of pepper pigments. Capsanthin is a powerful and safely natural coloring agent: these days it is used as a coloring in sausages, cheeses, fruit gelatins, drugs, and cosmetics. Included in chicken feed, it turns chicken skin yellow and attractively darkens the color of egg yolks; fed to dairy cows, less successfully, it produces pinkish milks and butters. It spruces up plumage color in cage birds and, in zoos, improves the look of dingy flamingoes.

Along with the versatile capsanthin, red peppers contain at least five other carotenoid pigments and some breeds contain as many as thirty. The yellow and orange capsicums get their color from beta-carotene cucubitene, also a major pigment of carrots. Green peppers are capsanthin-less; brown peppers, such as the "chocolate-colored" Mexican Pasilla, simultaneously contain chlorophyll and capsanthin, a mix of red and green that produces a biological brown. (The Pasilla, dried, resembles a raisin; hence its name, from the Spanish *pasa,* "raisin.") Green and red peppers generally represent not two different breeds of pepper, but two different developmental stages of the same fruit. On the road from green to red, they also pass through an intermediate brown stage known as mulatto, which occurs as the immature green pepper begins to accumulate red pigments while still retaining chlorophyll. Ripening times vary from cultivar to cultivar, but in general the green fruits are ready for picking about 70 days after fruit-set, and the fully mature red peppers at about 130 days.

Peppers can contain six times as much vitamin C as oranges, the highest levels of vitamin C being found in the green fruit. Levels decrease with maturity, and one researcher has found evidence that as capsaicin goes up, C drops off, which argues that the hottest peppers aren't necessarily the healthiest.

The picked pepper is a repository of several nutritional goodies. Peppers bulge with vitamins C, A, E, and P, thiamine (B_1), riboflavin (B_2), and niacin (B_3). Vitamin C was first purified in 1928 by Hungarian biochemist Albert Szent-Györgyi from a rejected supper dish of sweet paprika peppers. The sweet pepper concoction may have been the most fortunate failed recipe in history. Szent-Györgyi, who had been struggling unsuccessfully with bovine adrenal glands, described it as a "treasure trove" of his new vitamin, for which he was to win the Nobel Prize for Physiology and Medicine nine years later in 1937. Peppers can contain six times as much vitamin C as oranges and, as early as the seventeenth century, were taken to sea by Spanish sailors, who may or may not have recognized their usefulness as a scurvy preventive. The highest levels of vitamin C are found in the green fruit. Levels decrease with maturity, and one researcher has found evidence that as capsaicin goes up, C drops off, which argues that the hottest peppers aren't necessarily the healthiest. Raw fresh fruits are the best C sources;

content diminishes about 30 percent in canned or cooked peppers and essentially vanishes altogether from dried peppers. One three-ounce sweet pepper is enough to provide an adult with his or her U.S. Recommended Daily Allowance of vitamin C, with a bit left over.

Vitamin P, or citrin, also discovered in capsicum peppers by Szent-Györgyi and co-worker Istvan Ruznyak, is a regulator of vascular permeability. Vitamin A is also known as retinol because it is essential for the operation of the retina of the eye. Among the earliest signs of vitamin A deficiency is a loss of night vision. Two vitamin A precursors, carotene—so-named because it was believed to occur at highest levels in carrots—and cryptoxanthin are found in green and yellow vegetables. While carrots are a good source of A precursors, ounce for ounce they seem to be no better than capsicums; instead, top of the line for those desiring to see in the dark, according to the vegetarian cookbook *Laurel's Kitchen,* is a cup of cooked lambsquarters.

Famous chili fanatics include Will Rogers, who routinely judged a town by the quality of its chili, and Jesse James, who passed up the bank in McKinney, Texas, because the town harbored his favorite chili parlor.

Along with all these banner dietary benefits, the capsicums, often considered a spice rather than a vegetable, are a culinary treat. Capsicums are staples of Hungarian and Mexican cuisines, of Indian curries, and of Cajun gumbos. In this country, perhaps the best-known use of capsicums is in chili, a dish named after its peppery prime ingredient and originally derived from the ancient Nahuatl word *chilli,* meaning red. The sixteenth-century Mexican Nahuatl-speakers apparently did not eat anything recognizable to the modern palate as chili, though they did, according to a boggled Spanish observer, consume "frog with green chillis, newt with yellow chilli, tadpoles with small chillis, maguey grubs with a sauce of small chillis [and] lobster with red chilli, tomatoes and ground squash seeds."

The origin of the modern dish is a matter of debate among chili fiends. Various theories attribute its source to the chile-laden pemmican of the Plains Indians, to the Canary Islanders who introduced cumin seed to San Antonio in the 1730s, to the chuckwagon cooks on the cattle trails, and to the *lavanderas,* or laundresses, who accompanied the Mexican army during the border battles of the 1830s and 40s and who, along with scrubbing shirts, stuffed the soldiers with a mix of goat meat, wild marjoram, and red chiles known as son-of-a-bitch stew. From any or all of the above, chili has evolved into a meal for all regions: Maine chili is made with shell beans, California chili with avocados and olives, Alaska chili with moosemeat, and Texas chili (which originated, claims a vociferous Ohio faction, in Cincinnati) with goat, skunk, or snake. Teddy Roosevelt's Rough Riders made it with beefsteak and may have consumed a pot or two before taking part in the famous charge up San Juan Hill. While Old Army Chili (this page) hardly seems to have been a gourmet delight, civilian chili fanciers have done their bit to further the glorious cause. Chili powder, the story goes, was invented by Willie Gebhardt, a Texas German from New Braunfels, in 1892. It was a mysterious mix of dried peppers, oregano, cumin, and garlic that became the pillar of the chili con carne canning business in San Antonio. Chili powder has also had a few successes on the side: sprinkled in socks, it's said to keep your feet warm in winter, and sprinkled in sheets, it's said to get rid of bedbugs. Modern science has not yet supported the claim of one World War II pilot that his malaria was cured by a pot of chili, but it's something to think about in a pinch.

Famous chili fanatics include Will Rogers, who routinely judged a town by the quality of its chili, and Jesse James, who passed up the bank in McKinney, Texas, because the town harbored his favorite chili parlor. Today chili lovers have grown in number to the point of forming international organizations, including the International Connoisseurs of Green and Red Chile and the Chile Appreciation Society International, complete with conferences, newsletters, and annual cook-off championships. The cook-offs, which offer substantial financial prizes, were sponsored in part last year by Tabasco, Montezuma Tequila, and Pepto Bismol.

The capsicum peppers also serve a more elevated internal function than that of simple food. Medicinal use of capsicums has a long history, dating back to the Mayas, who used them to treat asthma, coughs, and sore throats. A modern-day version of the Mayan

OLD ARMY CHILI

From Manual for Army Cooks, *War Department Document #18, 1896*

1 beefsteak (round)
1 Tbs. hot drippings
1 cup boiling water
2 Tbs. rice
2 large dried red chile pods
1 cup boiling water
flour, salt, and onion (optional)

Cut steak into small pieces. Put in frying pan with hot drippings, cup of hot water, and rice. Cover closely and cook slowly until tender. Remove seeds and parts of veins from chile pods. Cover with second cup of boiling water and let stand until cool. Then squeeze them in the hand until the water is thick and red. If not thick enough, add a little flour. Season with salt and a little onion, if desired. Pour sauce over meat-rice mixture and serve very hot.
(Recipe designed for the individual mess kit.)

sore-throat remedy, suggested by no less an authority than the American Medical Association, is a gargle of ten drops of Tabasco sauce in half a glass of warm water. A sixteenth-century Spanish priest noted that moderate consumption of peppers "helps and comforts the stomach for digestion," which it may indeed have done: recent research has shown that capsaicin boosts secretion of saliva and stomach acids, and increases peristaltic movements. There is also some evidence that it possesses antibacterial activity, which makes it useful in some instances as a food preservative and is a good reason for making hot, instead of mild, sausage. It is featured in the deep-heat rubdown liniments used by athletes and other victims of aching muscles, and, non-medicinally, it's capsaicin that puts the zip in ginger ale and ginger beer.

Like so many other of the New World foods, the capsicum peppers were suspected of aphrodisiac properties. The same sixteenth-century priest who plugged peppers for the digestion, warned against too much of a good thing, "as the use thereof is prejudiciall to the health of young folkes, chiefely to the soule, for that it provokes to lust. . . ." This hopeful prediction has not been borne out to date by modern science, but one 1938 study of the effect of capsaicin on the water flea, *Daphnia magna,* showed that the compound elicited "pronounced and continued excitatory movements of the male genital organ," perhaps a nice sign for the future.

The peppers that produce all these spectacular effects have changed very little since the days of the Aztec planters. The first scientifically improved capsicum cultivar didn't come along until 1921, when Fabian Garcia, then director of the New Mexico Agricultural Station, released his "New Mexico Number 9." Number 9, product of a series of crosses involving a couple of local southwestern peppers and some dozen strains of the chocolate-colored Mexican Pasilla, was notable for disease resistance, high yields, and readily peelable pods, and as such became Number One among commercial pepper growers until well into the 1950s. (In memory of this landmark feat, Garcia's birthday, January 19, is celebrated annually by the International Connoisseurs of Green and Red Chile with feasts and fireworks.) Number 9 was replaced in 1959 by an earlier-ripening version, New Mexico Number 6–4, still the commercial best bet for red or green processed peppers. The scientific pepper really came into its own, however, in the 1960s and 70s, with the establishment of New Mexico State University's famed chile research center in Las Cruces. From NMSU have come innumerable bigger and better pepper cultivars, among them Roy Nakayama's whopping Numex Big Jim, fruits of which average eight inches, with champions up to a foot long.

In this country, the top pepper producers are California and Florida, who between them pump out nearly seven hundred thousand tons of peppers each year. Most of these are of the species *C. annuum* var. *annuum.* Top cultivars are the bite-less bell peppers, which make up about 65 percent of the U.S. crop; the Anaheim peppers, on the order of Garcia's Number 9 (14.4 percent); the pimiento peppers (5.3 percent); and the popular jalapeños (4.4 percent). The jalapeño, which takes its name from the ancient market of Xalapa in the Mexican state of Veracruz, has had a more impressive career than most: in November 1982, it became the first capsicum in space, when astronaut Bill Lenoir took a few on board the spaceship *Columbia.* The jalapeño appears in a vast array of cultivars, at least two of which, the Early Jalapeño and the American Jalapeño, originated in the United States. Chances are if you buy jalapeños in this country, you're getting one of the two. No self-respecting Mexican, rumor has it, would touch either of them with a ten-foot pole.

Bell peppers were developed by the Indians of Central America well before the arrival of Columbus. When the name *bell,* presumably referring to the pepper's shape, was first adopted is a mystery, though certainly among the earliest uses was by the pirate-surgeon Lionel Wafer in 1681. Wafer, physician on an English privateer, was ungratefully marooned by his shipmates in Panama after being wounded by a gunpowder explosion. He was taken in by a tribe of local Indians, about whom he subsequently wrote a book. They ate, he said, "two sorts of Pepper, the one called Bell-Pepper, the other Bird-Pepper," both in "great

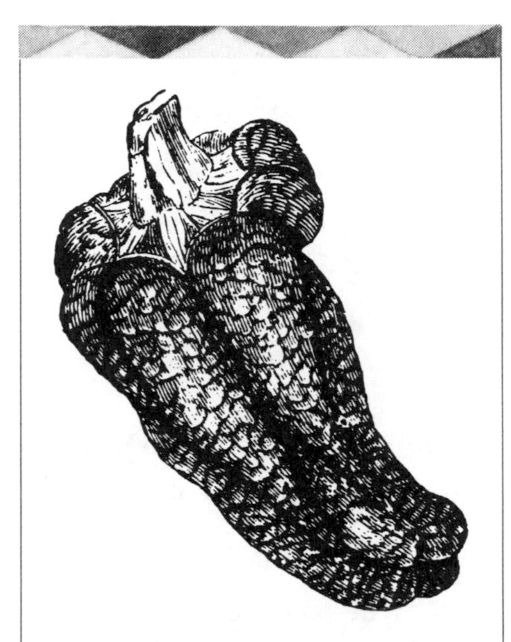

He was taken in by a tribe of local Indians, about whom he subsequently wrote a book. They ate, he said, "two sorts of Pepper, the one called Bell-Pepper, the other Bird-Pepper," both in "great quantities."

quantities." The most popular bell-type cultivar today is a breed called California Wonder, a blocky, thick-walled fruit introduced to gardens in 1928. The principal flavor component of the sweet bell pepper is 2-methoxy 3-isobutyl pyrazine, abbreviated MIP, which is also the compound that gives Cabernet Sauvignon its distinctive bouquet.

Anaheim peppers are the long green and red chiles so appreciated by the International Connoisseurs. Legend has it that their original ancestor was brought to New Mexico from Mexico in 1597 by Don Juan de Oñate, the founder and first governor of Santa Fe. Some three hundred years later, a visiting Californian named Emilio Ortega acquired some descendants of Oñate's peppers and with them established a highly successful chile cannery in Anaheim, which is where the peppers (unfairly) picked up their common name. Anaheim peppers, in the immature green stage, are the chiles of chile rellenos; in the mature red stage, dried, they are used in the preparation of chili powder and paprika.

Pimiento is an attractive and tasty pepper that, regrettably, is known to most of us solely as a stuffing for green olives.

While green chiles are usually used fresh, red chiles, in general, are dried. Perhaps the most common of the dried red peppers is the Ancho, a large, relatively mild capsicum popular in Mexico. Green, it is known as the Poblano, after the valley of Puebla south of Mexico City, where these peppers were first cultivated in pre-Columbian days. It is not grown to any extent in the United States. The easiest and most primitive—of drying methods is simply to let the fruit hang on the plant, an excellent technique in that it best preserves original pepper color. The modern pepper industry, however, reluctant to leave well enough alone, often dehydrates the harvested crop en masse in artificially heated drying tunnels. An exception to the dried-red rule is the sweet red pimiento pepper, a large, roughly heart-shaped fruit re-introduced to this hemisphere from Spain in 1911. It's an attractive and tasty pepper that, regrettably, is known to most of us solely as a stuffing for green olives.

Peppers were never a popular crop in colonial America, though the indefatigable Thomas Jefferson grew a few (imported from Mexico) at Monticello and George Washington planted some "bird peppers" at Mount Vernon, along with equally experimental royal palmettos, sandbox trees, physic nuts, and Guinea grass. (Washington also had a go at pistachio nuts, which should not have made it in Virginia.) By the latter half of the nineteenth century, however, peppers had trickled down to the general gardening public. Burpee's 1888 *Farm Annual* offered some twenty varieties, in green, red, and yellow, including the Celestial, the Red Squash (of Massachusetts origin and "handsome appearance"), the Spanish Monstrous (six to eight inches long), the Red Chili ("the best for pepper sauce"), the Long Yellow, and the Cranberry, said to look like one. Today there are over one hundred sweet pepper cultivars on the market and over half as many hot—which, with the annual new additions, should be enough to satisfy even the pickiest International Connoisseur.

A final note on peppers is a mathematical one: If you've spent any time mulling over the problem of the number of pickled peppers in Peter Piper's peck, the answer, in jalapeños, is something around 225. It's anybody's guess, however, how many seashells she sold down by the seashore.

POTATOES

Of the two thousand or so species in the bulging genus *Solanum,* about 170 are tuber-bearers. Of the tuber-bearers, only eight are routinely cultivated and eaten by people, and most of these have stuck pretty much close to home in the Andes of Peru. Only one has reached international stardom: *S. tuberosum,* commonly known as the potato. The potato probably originated in Peru, where indications are that it was domesticated over six thousand years ago by high-altitude-dwelling ancestors of the Incas. To the original planters it must have been a godsend, since not much else grows readily in the Andean high sierra. While corn wimpishly peters out around eleven thousand feet, potatoes proliferate undaunted up to fifteen thousand, which means, should you be so inclined, you could put in a profitable potato patch halfway up Everest.

The original potatoes were small by modern standards—plum- or even peanut-sized—and the Incas wolfed them down along with dishes of llama, guinea pig, squash, beans, and, closer to sea level, tomatoes, peppers, and avocados. Ancient Andean potato cuisine was dominated by *chuño,* an unappetizing form of processed potato made by freezing, thawing, and stamping on the unfortunate tubers repeatedly until they were reduced to a blackened desiccated mass. This preparation had to be reconstituted with water before eating and, as such, can be viewed as a sort of primordial instant mashed potatoes. Chuño, similarly, was noted for its superb keeping qualities.

The Peruvians, proud of their potatoes, immortalized them in pottery: archaeologists have unearthed potato-shaped funeral urns, potato-decorated cooking pots, and, for junior potato-eaters, potato-shaped whistles. The Quechua Indians, modern descendants of the Incas, have amassed over one thousand different names for potatoes, linguistic evidence of its immense regional importance. Our word *potato* perversely derives from none of them but from the Caribbean *batata,* which meant sweet potato, and was a mistake on the part of the Spaniards.

The Spaniards got their hands on the potato in the early sixteenth century, when Pizarro and company—out after gold and emeralds—stumbled upon them somewhere outside of Quito, Ecuador. Pizarro, who must have had an imaginative palate, described his vegetable find as "a tasty, mealy truffle" and, under the name *tartuffo,* the potato was introduced to Spain. From there, it meandered into Italy and France, where it was rejected by the general public on the grounds that the knobby, deep-eyed tubers resembled leprous hands and feet and were doubtless carriers of the disease. (Those few brave souls who tried it anyway compared it, puzzlingly, to chestnuts.) The English, despite much loose talk about the "Virginia potato," picked theirs up not in Virginia, but in Cartagena, Colombia, where Sir Francis Drake, after a profitable season of picking off Spanish treasure ships in the Caribbean, paused to lay in supplies for the long sea voyage home. His potato-stocked vessel then stopped off at Roanoke Island in Virginia to collect a handful of discouraged early colonists, and all returned to Mother England in 1586. (Those colonists who refused to accompany the potatoes home subsequently vanished in naggingly mysterious fashion, leaving behind them only a baby's shoe and the word *Croatan* carved on a tree trunk.) Samples of the leftover Colombian potatoes were passed on to herbalist John Gerard, who never did sort out where they came from. He thought them "mighty and nourishing" and pointed out that their regrettable tendency toward "windinesse" could be eliminated by eating them sopped in wine. By the 1633 edition of his *Herball,* potatoes rated a whole chapter of their own, still stubbornly titled "Of Potato's of Virginia." The Virginia potatoes—by some, admits Gerard, called Skyrrets of Peru—were designated Common, or Bastard, Potatoes, presumably to distinguish them from the genuine article, the sweet potato.

The sweet potato was discovered by Columbus

Queen Elizabeth's cooks, uneducated in the matter of potatoes, tossed out the lumpy-looking tubers and brought to the royal table a dish of boiled stems and leaves, which promptly made everyone deathly ill.

on his second trip to the New World and sent back to Spain in 1494 along with a number of unhappy Indians, sixty parrots, and three gold nuggets. The sweet potato, *Ipomoea batata,* is a relative of the morning-glory and botanically a whole different ballgame from its common namesake. The scientific name comes from the Greek *ips* ("worm") and *homoios* ("like"), since Carolus Linnaeus—the eighteenth-century Swedish botanist, famed for his system of plant classification—thought the twining vines looked unpleasantly like worms. Beneath these wormish vines, the roots accumulate stored food and swell to form sweet potatoes. In contrast, the common potato, for all its suggestive underground location, is not a root vegetable, but a tuber, the outgrowth of an underground stem, or *stolon.* (The result, however, is the same: you have to dig it up.) Sweet potatoes were considered delicacies through the sixteenth and early seventeenth centuries, on par with such exotic goodies as oranges and dates. They were particular favorites of Henry VIII, who, judging by his portraits, was a first-class food fancier. Henry preferred his potatoes baked in pies, and a surviving late sixteenth-century recipe calls for combining potatoes with quinces, dates, egg yolks, the brains of three or four cock sparrows, sugar, rose water, spices, and a quart of wine. The sweet potato was also considered an aphrodisiac as well as a taste treat: when Shakespeare's Falstaff shouts "Let the sky rain potatoes!" in *The Merry Wives of Windsor,* he was hoping for *Ipomoea batata.*

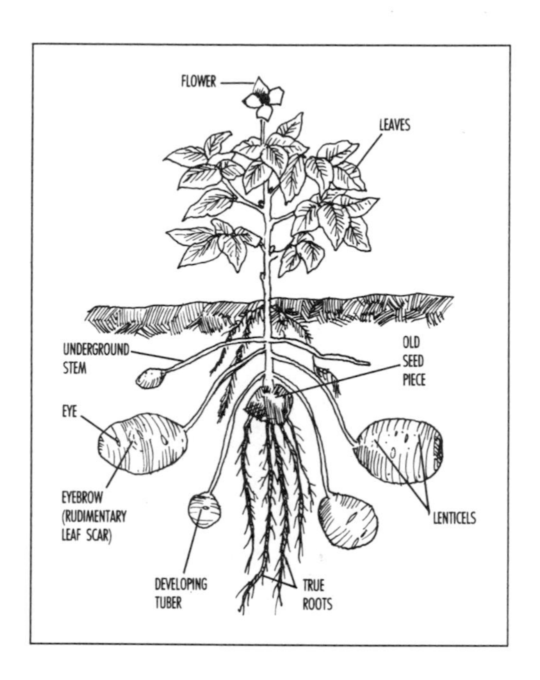

The common potato might have had better luck if English cooks had stuck to pie cuisine. However, it seems that they didn't, and the upsetting result did little for the potato's popular reputation. A few of Drake's potatoes, the story goes, were given to Sir Walter Raleigh, who planted them on his estate near Cork, Ireland, and later gallantly made a gift of potato plants to Queen Elizabeth I. Queen Elizabeth's cooks, uneducated in the matter of potatoes, tossed out the lumpy-looking tubers and brought to the royal table a dish of boiled stems and leaves, which promptly made everyone deathly ill. Potatoes, understandably, were banned from court and it was some centuries before they managed to live down their public image.

The true culprits in the royal banquet disaster were the poisonous potato alkaloids solanine and chaconine, present in highest quantity in the stems and leaves of the plants. Tiny amounts of these compounds are also present in the tubers, where, under normal circumstances, they contribute positively to the potato's overall taste. Under certain conditions, however, these compounds can accumulate to the point of toxicity. Alkaloid production in tubers is turned on by exposure to light or to extremely cold or hot storage temperatures. Luckily for the unwary, light also stimulates the production of chlorophyll, which means that dangerous tubers are often green—plus, the alkaloids, rather than spreading themselves promiscuously throughout the tuber, tend to concentrate in a thin layer at the surface and thus can be disposed of by vigorous peeling. The sprouts—infant stems—are particularly heavy in solanine and chaconine, and should be mercilessly excised.

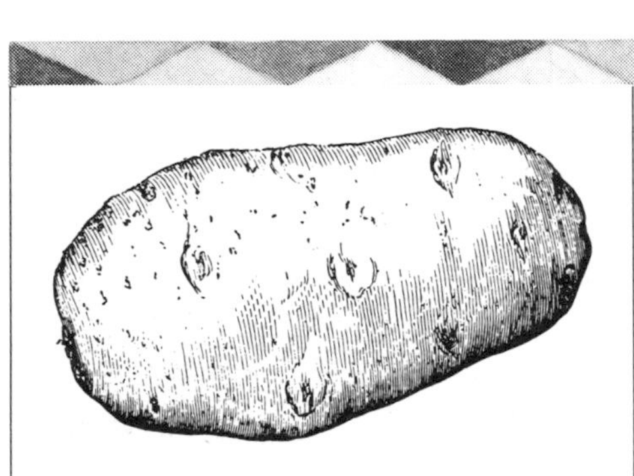

Nervous Presbyterian ministers in Scotland forbade it on the grounds that nobody mentioned it in the Bible. There were even hints that the potato may have been the Forbidden Fruit in the Garden of Eden.

While Elizabeth I's unfortunate helping of potato stems slowed the acceptance of the tuber in some quarters, others rejected it for more creative reasons of their own. Nervous Presbyterian ministers in Scotland forbade it on the grounds that nobody mentioned it in the Bible. There were even hints that it may have been the Forbidden Fruit in the Garden of Eden, which leaves us with an appealing vision of Eve and the snake grubbing about with a shovel. (Modern scholars, however, think the forbidden fruit was an apricot.) The Dutch introduced the potato to Japan in the early seventeenth century, where it was relegated to use as cattle fodder until Commodore Perry talked the Emperor into trying a few in 1854. Peter the Great acquired potatoes on a visit to Holland in 1697 and brought them home to Russia as a treat for imperial banquets; in the next century, his ungrateful peasants spurned them as "the Devil's apples." Apprentices in colonial America refused to eat them, claiming that potato-eating might shorten their lives, and as late as the mid-nineteenth century, many thought potatoes fit only for livestock. A contemporary *Farmer's Manual* suggested they be planted near the hog pens, the better for convenient feeding.

Over the years, potatoes have been blamed for

rickets, scrofula, leprosy, and syphilis, and as late as 1904, Célestine Eustis in *Cooking in Old Créole Days* warned prospective Cajuns that "water in which vegetables have been boiled can be used in cooking, except potato water and cucumber water. They have been known to poison a dog." More insidious was the potato's proposed role in moral decay: in the late nineteenth century, Reverend Richard Sewall accused the potato of leading to wantonness in housewives, since its preparation required little time and effort, thus leaving female hands idle and primed to do the Devil's work.

While little early enthusiasm was shown for personal potato-eating, many were eager to foist a potato diet off on others. It was seen almost immediately by wealthy civilians as a handy solution to the perennial food problems of the poor, the army, the jails, and the insane asylums. In 1664, John Forster, Gent., produced his magnum opus "Englands Happiness Increased, or a Sure and Easie Remedie against all succeeding Dear Years; by a Plantation of the Roots called Potatoes . . . Invented and Published for the Good of the Poorer Sort." The Easie Remedie relied on using potatoes to make bread, and Forster's recipe starts out with a bushel of boiled potatoes, to be placed in a "wier sieve" and smashed with an "Iron Truel," a process that never really caught on. In 1683 the Royal Society of London began pitching potatoes as a famine relief crop. The Irish, a Poorer Sort if there ever was one, were growing potatoes as a staple by the late seventeenth century, and a typical peasant family (man, wife, and four children) was said to consume 252 pounds of the tubers each week, eked out with 40 pounds of oatmeal, a little milk, and an occasional salted herring. Equally early potato-eaters were the Germans, who have the distinction of publishing the first-known potato recipes, in *Ein Neu Kochbuch* (1581) printed on the august press of Johann Gutenberg. Still, potato acceptance by the average German was not without a struggle. The winning trick may have been the "Brandenburg Potato Paper," an uncompromising late seventeenth-century edict issued by Emperor Frederick Wilhelm following a series of disastrous Prussian crop failures. A masterpiece of

The Irish were growing potatoes as a staple by the late seventeenth century, and a typical peasant family (man, wife, and four children) was said to consume 252 pounds of the tubers each week, eked out with 40 pounds of oatmeal, a little milk, and an occasional salted herring.

simplicity, it ordered all peasants to plant potatoes or have their noses and ears cut off. Frederick Wilhelm's famous grandson, Frederick the Great, made the point more subtly, by publicly eating potatoes on the balcony of the Imperial Palace, and by distributing seed potatoes to landowners. By then, however, potatoes were an established crop, enough so to allow Frederick to embroil his subjects in the Potato War (*Kartoffelkrieg*) of 1778–79, more formally known as the War of Bavarian Succession. The war, the latest in a century's-worth of ill-mannered bickering between Prussia and Austria, acquired its vegetable nickname because it consisted largely of opposing forces destroying each other's potato fields.

The potato arrived in colonial America, it is believed, in 1622, when the governor of Bermuda sent two chestsful as a present to the governor of Virginia. Nothing much seems to have been done with them; nor with the fifteen tons of potatoes that were delivered, along with assorted oranges and lemons, to the Massachusetts Bay Colony in 1636. Potatoes (still imported) appeared on a Harvard dinner menu celebrating the installation of a new president in 1707, though in considerably smaller quantities than the brandy, beer, Madeira, and "green wine." The first substantial patch of homegrown potatoes in the colonies seems to have been planted in 1719 near Londonderry (now Derry), New Hampshire, by a newly arrived batch of Scotch-Irish settlers. In the hands of the Scotch-Irish, the "Irish" potato flourished and spread to adjacent settlements, reaching Connecticut in 1720 and Rhode Island in 1735. New England, ever with an eye to the main chance, was growing enough to export by 1745. Still, the eighteenth-century potato was viewed in most circles as something you ate only when everything else was exhausted. Potato consumption increased during the lean years of the Revolutionary War, and John Adams, who viewed this as a hardship, wrote in a patriotic letter home to Abigail, "Let us eat potatoes and drink water . . . rather than submit." She responded feistily that they could probably do as well on whortleberries and cow's milk.

John Adams's opinion of the potato was firmly seconded by the eighteenth-century French. Described in 1749 by naturalist Raoul Combes as "the worst of all vegetables," the potato only began to make public headway in France in the 1770s, under the auspices of Antoine-Augustin Parmentier. Parmentier's enthusiasm was such that his name became almost synonymous with potato—there was, in fact, a move to rename the tuber *parmentière* in his honor, which failed. (He was, however, preserved for posterity in Potage Parmentier, a rich potato soup.) Parmentier, originally a military pharmacist, adopted the potato following a stint as a prisoner in Germany during the Seven Years' War. He was fed almost exclusively on potatoes during his captivity and must have thrived on them, since he returned home convinced that the tubers had unplumbed possibilities.

In 1771, he got his chance to present these possibilities to the public: in that year, following a severe crop failure, the Academy of Besançon offered a cash prize to whoever came up with the best "study of food substances capable of reducing the calamities of famine." Parmentier won hands down with his "Inquiry Into Nourishing Vegetables That in Times of Necessity Could Be Substituted for Ordinary Food." Foremost among the Nourishing Vegetables was the potato, weakly followed by the acorn, the horse chestnut, and the roots of irises or gladioli. Prize or no, the potato did not immediately leap to prominence, and Parmentier was to spend the next decades promoting his chosen vegetable. Part of the problem may have been his emphasis, which was less on the potato as vegetable than on potato starch as a substitute for wheat flour in baking. This never worked all that well, though Voltaire, in one of his back-to-nature periods, tried mixing potato starch fifty-fifty with wheat flour and managed to turn out "a very savorous bread." Outside of the kitchen, potato starch did achieve a certain popularity as a wig whitener.

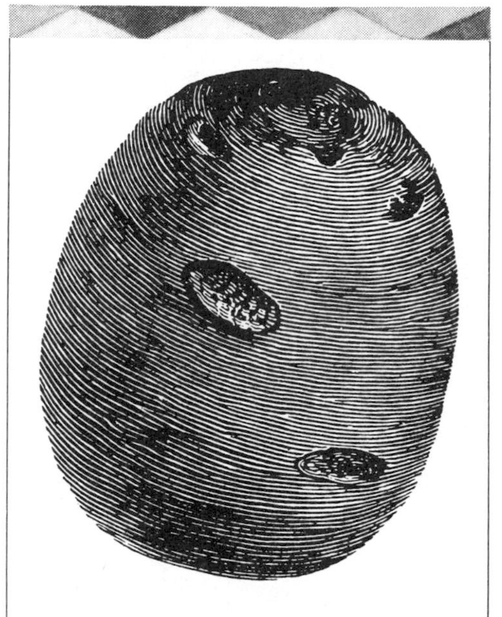

The king tucked a flower in his lapel, Marie Antoinette stuck one in her coiffeur, and the potato, socially, was made.

Success came on the King's thirty-first birthday, August 23, 1785, when Parmentier foxily presented Louis XVI with a bouquet of potato flowers. The King tucked a flower in his lapel, Marie Antoinette stuck one in her coiffeur, and the potato, socially, was made. During his time in the royal sun, Parmentier supervised the preparation of a totally tuberous court dinner, featuring some twenty potato dishes, from potato soup to post-prandial potato liqueur. Benjamin Franklin, unconfirmed rumor has it, was one of the guests. While the potato soon declined from these dizzying heights of favor, it was established in France from the late eighteenth century onward as a useful and reputable field crop. In memory of his service to humankind, potatoes used to be—perhaps still are—planted annually on Parmentier's grave.

Parmentier's potato, in the very teeth of its vociferous detractors, still managed to accumulate a substantial history as an effective medicinal. In one form or another, it was considered a cure for gout, lumbago, rheumatism, sore throat, sunburn, frostbite, drunkenness, black eyes, temper tantrums, sprains, sciatica, warts, toothaches, and (in horses) thrush. In 1824, *The Family Oracle of Health,* under the heading "Beauty Training for Ladies," decreed that hopeful females should breakfast on biscuit and beefsteak and dine on broiled chicken and potatoes. Such a diet, according to Dr. Martyn Payne of New York, would also successfully fend off cholera.

While Dr. Payne stood on scientifically shaky ground, the potato in a nutritional sense is a vegetable gold mine. One medium-sized potato contains 3 grams of protein, 2.7 grams of dietary fiber, and 23 grams of carbohydrate. Each potato also contains about half the adult Recommended Daily Allowance of vitamin C—the Spaniards used potatoes as antiscorbutics on board the treasure galleons—plus reasonable amounts of niacin, riboflavin, and thiamine. The assessed biological value of the potato—the index of the amount of nitrogen absorbed from a given food that is actually used by the body for maintenance and growth—is seventy-three, as compared to ninety-six for the egg (considered the most biologically complete of foods), seventy-two for the soybean, and fifty-four for corn-on-the-cob. Potatoes contain more potassium than bananas, but practically no sodium, and next to nothing in the way of fat.

Despite all their pudgy press, potatoes are far from fattening. Consisting of about 80 percent water, they possess a mere one hundred calories or so per tuber, as opposed to eighty for the average apple. The problem with the potato, dieticians point out, lies in what we pile on it. Few of us after all are given to slathering apples with butter (thirty-six calories per pat) or sour cream (twenty-six calories per tablespoon).

Most of the potato's carbohydrates are in the form of complex chains rather than simple sugars, and most of those chains (66 percent of potato dry weight) consist of starch. The quickie test for starch, as you might remember from Biology 101, is to dunk the questionable material in iodine; if it turns black, it's starch. Perhaps the most creative use of the starch test in history was that of master criminal John Dillinger, who carved a pistol out of a potato, stained it with iodine, and used it to escape from jail. In these less spectacular days, complex carbohyrates have been shown to lower the fat content of the blood, thus reducing the risk of arteriosclerosis. They have also been recommended in cancer-preventive diets. It has been shown that diets high in fat and low in fiber and complex carbohydrates—a regrettable American phenomenon—are linked to the incidence of colon cancer. To reverse this unfortunate trend, the National Institute of Health suggests a reduction in dietary meats, eggs, and dairy products, and increased consumption of grains, fruits, and vegetables.

The Irish Potato Famine of 1846–1848 has been cited as Europe's worst disaster since the devastating passage of the Black Death in 1348–50.

By the early nineteenth century, the Irish had been healthily living off potatoes for over two hundred years. While this would seem to speak well of the potato, in practice it served only to speak ill of the Irish. The two-faced English, who had shoved the potato down Irish throats in the first place, announced in an 1829 issue of the London *Times* that potatoes were fine for the wealthy, but if the lower classes ate too many of them, Great Britain would become "a nation of miserable, turbulent drunkards"—just like, the implication was, their Gaelic next-door neighbor. In Germany, the philosopher Friedrich Nietzsche stated disagreeably, "A diet that consists predominantly of rice leads to the use of opium, just as a diet which consists predominantly of potatoes leads to the use of liquor." Furthermore, the outrageous size of the average Irish family was taken by some as proof that the potato was indeed a powerful aphrodisiac. The Irish themselves naturally viewed matters somewhat differently. Two things are too serious to joke about, an old Irish saying goes: marriage and potatoes. The potato-dependent Irish referred to their tubers—seriously—as the Apples of Life.

The Irish themselves brought along the name *spud,* which is said to come via Gaelic and Cockney from the word *spade.*

A primary advantage of the Irish potato, in a country where farmland was at a premium, was its productivity. Potatoes yield a fivefold greater crop per unit land area than either wheat or corn, and by the mid-nineteenth century were comfortably supporting an Irish population of nine million, each of whom consumed an average ten pounds of potatoes per day. The cloud in this happy picture was the ever-present fear of potato crop failure and ensuing famine, a disaster that struck Ireland at least twenty-four times between 1728 and the now world-famous Great Hunger. The Irish Potato Famine of 1846–48 has been cited as Europe's worst disaster since the devastating passage of the Black Death in 1348–50. Attempts to cope with crop failure of such scope were futile. British Prime Minister Sir Robert Peel (after whom the London cops are called bobbies and the Irish ones are called *peelers*) tried to alleviate the situation by shipping in half a million dollars' worth of Indian corn from the United States—unsuccessfully; the Irish, unable to cook or eat it, referred to the British offering as "Peel's brimstone." Before the famine ran its course, it killed off 1½ million people in Ireland and forced another 1½ million to emigrate to the United States, England, and Australia.

The emigrants brought with them their predilection for potatoes, which in America were soon nicknamed mickeys or murphys in honor of their prime consumers. The Irish themselves brought along the name *spud,* which is said to come via Gaelic and Cockney from the word *spade.* An alternative explanation claims it derives from the acronym for the Society for the Prevention of Unsatisfactory Diets, an anti-potato organization active in late seventeenth-century England. The Irish and their potatoes were so closely linked in the popular mind that the predominantly Irish Boston police were known for a period of time as the Blue Potatoes. (Less kindly, the sudden wave of Irish arrivals spawned considerable anti-Irish slang, including the expressions *Irish buggy* ("wheelbarrow"), *Irish confetti* ("bricks"), *Irish nightingale* ("bullfrog"), and *Irish spoon* ("shovel").

While their reception in the new country may not have been all that could be desired, it was still better than staying at home to starve. While the desperate victims blamed the 1846–48 crop failures on everything from steam locomotives to volcanic eruptions, the evil genius behind the Great Hunger was a fungus, *Phytophthora infestans,* and the disease that did in the Irish potato crop is commonly known as late blight. Usually it affects the plant leaves first, then fungal spores wash into the soil and eventually infiltrate and destroy the underground tubers. By no means limited to Ireland, *P. infestans* has made itself felt wherever potatoes are grown, from the Andes to the Himalayas. The 1846 round of late blight also played havoc among American potato growers, dropping production over 40 percent. Though economically painful, this was less catastrophic in the United States, where the general diet depended less on potatoes and more on meat and gravy.

One positive result of the Potato Famine, if such a thing is possible, was the subsequent upsurge of interest in breeding new and blight-resistant potato varieties. The decimated Irish crop was descended from only one or two ancestral Peruvian potato specimens and thus consisted of essentially identical—and identically blight-sensitive—plants. Most seem to have belonged to a variety descriptively known as the Lumper. Named potato cultivars showed up sometime in the mid-1700s; before that, varieties were vaguely differentiated on the basis of color and shape. A list of Irish potato types dating to 1755 includes the "rough coat, red coat, flat white and long white" plus a small reddish number called the "Spanish potato" dismissed as fit only for cattle and swine.

Thomas Jefferson, usually more precise, planted "round potatoes" at Monticello in the last quarter of the century, and, in 1796, Amelia Simmons (who describes herself as "a poor solitary orphan") discussed five varieties in her *American Cookery,* the first published American cookbook. Amelia's pick was the How's Potatoe, a smooth-skinned potato "most mealy and richest flavor'd," followed by the yellow rusticoat (rusty-coated), then the red and the red rusticoat (both "tolerable"), and finally the yellow Spanish. Named late-eighteenth-century potatoes included the Howard, the Irish Apple, the Manly, the White Kidney, the Ox Noble, the Ashleaf, and the Lapstone Kidney, as well as the hapless Lumper. Bernard M'Mahon's 1806 *American Gardener's Calendar* mentions only one variety of potato in his list of sixty-seven "Esculent Vegetables," but by mid-century at least one hundred varieties were available, and multiplying. Popular were English Whites and Biscuits (both round), La Plata (a long red), Chenango, and Pennsylvania Blue.

Today production of potatoes tops 290 million tons a year, putting the potato sixth on the list of the world's major crops, behind sugar cane, wheat, rice, maize, and sugar beets.

The catastrophic passage of *Phytophthora infestans* had a substantial impact on the types of potatoes grown, and breeders were egged on by such incentives as the offering of a ten-thousand-dollar prize by the Commonwealth of Massachusetts to the person to discover "a sure and practical remedy for the Potato Rot." The best of the proposed remedies was a new potato known as the Garnet Chili, developed in 1853 by Reverend Chauncey Goodrich of Utica, New York, from a wild South American variety. From the Garnet Chili, Albert Bresee of Hubbardton, Vermont, produced the Early Rose, an introduction of the 1860s that rapidly became America's top potato. The Early Rose gave way in the 1870s to the Burbank potato, developed in 1873 by twenty-three-year-old Luther Burbank from a seedball stumbled across in his mother's Massachusetts garden. Burbank sold his landmark potato for $150 to nurseryman J. J. H. Gregory and used the proceeds to move to California. There, he settled down in Santa Rosa, and went on to create a white blackberry, a stoneless cherry, and a spineless cactus, plus a grand total of seventy-eight new fruits, nine new vegetables, eight new nuts, and several hundred varieties of ornamentals, including the Shasta daisy. His potato is the ancestor of the Russet Burbank, the potato that made Idaho famous.

By the turn of the century, potato varieties numbered in the thousands. The seed house of Vilmorin-Andrieux, announcing despairingly that "the number of the varieties of the Potato is prodigious," listed a mere 135, including 31 French varieties; 18 German; 19 American, including the jumbo White Elephant; and 25 English, among them the quintessentially British Rector of Woodstock and Vicar of Laleham. Today production of potatoes tops 290 million tons a year, putting the potato sixth on the list of the world's major crops, behind sugar cane, wheat, rice, maize, and sugar beets. (The sweet potato, at one hundred million annual tons, ranks ninth.) The U.S.S.R. is the world's top potato producer, turning out about a quarter of the total crop, most of which goes into vodka or livestock feed. Next in line are China, Poland, Germany, the United States, and Ireland. The U.S. accounts for about 14½ million tons of potatoes a year, over half (8½ million tons) from Idaho. Next-best national producers are Washington, North Dakota, and Maine, where in the heart of potato-laden Aroostook County, a Potato Blossom Festival is held each year, featuring mashed-potato wrestling, a parade, and the crowning of a Potato Blossom Queen.

While there are now over five thousand varieties of potatoes—with new and improved hybrids popping out of research laboratories daily—80 percent of the American crop derives from just six cultivars. Nearly half of that crop is Russet Burbank, the American generic potato, from which McDonald's makes their french fries. Other biggies are Kennebec, Katahdin, Norgold, Norchip, and Superior. Such lack of genetic originality is a matter of considerable concern among agriculturalists and related interest groups, who emphasize that uniform crops possess dangerously uniform susceptibility to disease. A case in point is the 1970 corn crop failure, when 50 percent of America's genetically identical cornfields fell before southern corn blight. Foremost in the battle for potato diversity is the International Potato Center (Centro Internacional de la Papa), appropriately based in Lima, Peru, ancestral home of the potato. The Center, staffed by seventy scientists from twenty different countries, possesses the world's largest collection of potato cultivars—about 5,500 at last count—and simultaneously maintains this existing pool of priceless genetic material, while attempting to develop new and better potato varieties.

Potatoes are usually propagated using seed potatoes—chunks of parent potato containing an "eye"—

an unexciting asexual process that yields offspring genetically identical to the parent plant. True potato seed is contained in the potato berry, a small green tomato-like fruit that usually gets tossed unappreciatively on the compost heap. True seed is tiny. There are fifty thousand potato seeds in an ounce, enough to plant an acre's-worth of potatoes, as opposed to a bulky required sixteen hundred pounds of seed potatoes. Potato seed is relatively disease-free, as compared to the tubers, which are notorious carriers of viruses. And perhaps most important, true potato seed has the potential for introducing genetic diversity into the largely all-of-a-kind national potato crop.

Unfortunately, such diversity is a two-sided coin: since potato seed is a genetically mixed bag, developing varieties that breed true is no scientific Sunday picnic. Most potatoes are tetraploid—containing quadruple sets of twelve chromosomes each—and such complex hybrids are difficult to sort out in even the most dedicated laboratory. A good start along these lines was the creation of the Explorer potato, a seed-reproduced cultivar developed after eight years of research effort by the Pan-American Seed Company in Chicago.

Explorer seeds yield small, round, brown-skinned potatoes, but, according to plant specialists, not nearly enough of them to satisfy the commercial growers. A reasonable commercial expectation is three or four pounds of potatoes per hill; Explorer delivers about one-eighth of that. Quantity counts: the average American eats about 115 pounds of potatoes a year. About half of those are fresh potatoes, half processed: that is, frozen, dehydrated, french-fried, or chipped. Processing results in unavoidable vitamin loss, but that doesn't seem to hold American consumers back any. In this country, five billion pounds of potatoes a year go to make french fries.

The Moon's Lake House chef, an American Indian named George Crum, ran afoul of a cantankerous customer—embellished versions claim it was Cornelius Vanderbilt—who kept sending his fried potatoes back to the kitchen, complaining that they were too thick. Crum, driven to the wall, finally sliced his potatoes paper-thin and served up the fried result.

Thomas Jefferson, a perennial food freak, encountered "French" fried potatoes while serving as ambassador to France in the 1780s and became fond enough of them to offer fries to guests at Monticello once he returned home. In spite of this elite introduction, french fries didn't catch the public fancy until the 1870s and weren't really common until the twentieth century. They were known quite formally as "French fried potatoes" until the 1920s, when the name was shortened to "French frieds"; then a decade later it was truncated even further to the now-familiar "french fries." Most french fries today are Russet Burbanks, vaguely rectangular potatoes eminently suitable for dissection into squared-off strips, but a vocal Vermont contingent holds that the prime fries are made—fresh—from Green Mountain potatoes, a cultivar developed by Vermont growers Brownell, Rand, and Alexander in 1885. French fries seem to be one of the few products in which the United States has managed to surpass Japan; American processors have cornered 90 percent of the Japanese french-fry market.

In Great Britain, french fries are known as chips, as in "fish and chips," while potato chips are known as crisps, presumably because they are. However, potato chips, like ice-cream cones, hula hoops, and frisbees, are an American invention. The story goes that potato chips first came to light in the late 1800s at the Moon's Lake House in Saratoga Springs, a then-fashionable upstate New York spa. The House chef, an American Indian named George Crum, ran afoul of a cantankerous customer—embellished versions claim it was Cornelius Vanderbilt—who kept sending his fried potatoes back to the kitchen, complaining that they were too thick. Crum, driven to the wall, finally sliced his potatoes paper-thin and served up the fried result. The potatoes were a vast success and for years afterward, dubbed "Saratoga chips," were a specialty of the Moon's Lake House, stuffed into paper cornucopias made by the owner's wife.

Potato chips these days are America's top snack, out-selling popcorn, pretzels, peanuts, and tortilla chips, to the tune of three billion dollars annually. Chips are also an international favorite, variously flavored with seaweed for the Japanese market, paprika for Berliners, and curry for the chip-eaters of New Delhi. Modern chips are generally skinnier than Crum's originals: most are around fifty to sixty thousandths of an inch thick, with more substantial varieties such as Frito-Lay's Ruffles (with ridges) about twice that. They're not hand-sliced anymore either, but are whipped out by

SARATOGA POTATOES

From Mary F. Henderson, Practical Cooking, and Dinner Giving, *1878*

♦ It requires a little plane, or potato or cabbage cutter, to cut these potatoes. Two or three fine, large potatoes (ripe new ones are preferable) are selected and pared. They are cut, by rubbing them over the plane, into slices as thin or thinner than a wafer. These are placed for a few moments in ice, or very cold water, to become chilled. Boiling lard is now tested, to see if it is of the proper temperature. The slices must color quickly; but the fat must not be so hot as to give them a dark color.

Place a salt-box on the hearth; also a dish to receive the cooked potatoes at the side; a tin plate and perforated ladle should be at hand also. Now throw, separately, five or six slices of cold potato into the hot lard; keep them separated by means of the ladle until they are of a delicate color; skim them out into the tin plate; sprinkle over some salt, and push them on the dish. Now pour back any grease that is on the tin plate into the kettle, and fry five or six slices at a time until enough are cooked. Two potatoes fried will make a large dishful.

mechanically flinging potatoes against a spinning drum stubbed with knife blades. The final chips are extensively tested for such qualities as fracture pattern—do they break into sharp little points when bitten?—and mouth clearance—how long do they take to chew? Varieties most frequently used as "chipping" potatoes are Katahdin and Norchip.

Dehydrated potatoes, of which Americans ate eleven pounds apiece in 1982, first made it big in World War II, when they were mercilessly fed to the U.S. Army. Instant-potato proponents cite the average 20 percent nutritional loss that results from the amateur peeling of fresh potatoes in the home or institutional kitchen—easy to do, since one-third of the potato's nutrients are squeezed into a thin band called the cortex located just beneath the peel. It shows up as a darker border around the rim of potato chips. Instant potatoes are also said to be cheaper than genuine mashed potatoes. My personal feeling, however, is that potatoes should not have the consistency of Cream of Wheat. Perhaps the best use of dehydrated potatoes to date was by assorted film makers who used the flakes in Christmas movies to imitate snow.

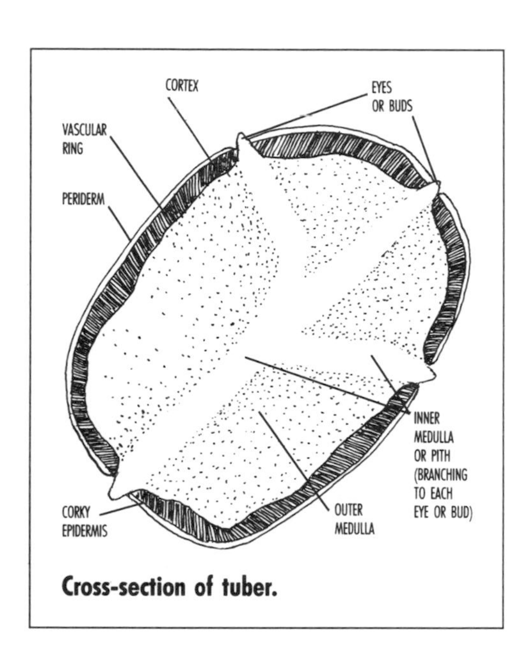

Cross-section of tuber.

More appealing is the alcoholic potato, nowadays a prime ingredient of vodka. (Vodka itself considerably predates the European potato, originating in fourteenth-century Russia, where it was first made from wheat or rye.) Early Americans, though still leery of eating potatoes, were turning them into whiskey in the eighteenth century, and a mysterious Mackenzie, who published his *5,000 Receipts* in 1829, recommended potato wine. "Wine of considerable quality," says Mackenzie optimistically, "may be made from frosted potatoes, if not so much frosted as to have become soft and waterish." Would-be winemakers are instructed to smash up a Winchester bushel of potatoes with a wooden mallet, add hops, white ginger, water, yeast, and sugar, and let age for three months. Prospectors in Alaska during the Far Northern gold rush of the 1880s taught the natives to produce a distilled potato concoction called hoochinoo—later shortened to "hooch"—perhaps from such local potato varieties as Gold Coin, Extra

Early Triumph, and Extra Early Eureka.

Marilyn Monroe once posed in a potato sack, looking delectable, and doubtless giving new meaning to the term *hot potato,* which, since the 1920s, has meant a spectacular girl. For those lucky enough to have one, the National Potato Promotion Board, based in Denver, Colorado, has decreed that February is National Potato Lover's Month, nicely timed to overlap with Valentine's Day. If you'd really like to make points, there's even a potato cultivar known as Red Rose.

EGGPLANTS

The original eggplant, botanists believe, blossomed somewhere in south central Asia—possibly India—where its peculiar-looking fruits, bitter taste, and nasty thorns did little to recommend it to the primitive palate. Nonetheless, some hardy soul eventually domesticated it, and by the third century A.D. the Chinese were gingerly debating its dietary potential. Putting theory into practice was a long drawn-out process: the Oriental eggplant seems to have remained an object of philosophical speculation for another three hundred years, and when it finally made it to the table, it was eaten nervously. Like the notorious blowfish, the eggplant was thought safe only if prepared by trained cooks—its Chinese name, *ch'ieh-pzu,* in most pessimistic translation, means poison. However, as increasing numbers of diners cleaned their plates without lethal incident, eggplants, if not popular, became commonly acceptable.

In India, eggplant—somewhat disguised—figured in the traditional curries, spicy sauces concocted to add pizzazz to rice or chapatis. A typical early curry was soup-like—a European observer described it as a broth—and contained a chopped eggplant plus onions and lentils, cooked in *ghi* (clarified butter), flavored with cardamom, coriander, cumin, turmeric, white pepper, and mustard seed, and diluted with coconut milk. Though respectable, this was far from the culinary stardom that the eggplant attained once it reached the Middle East. In the Middle East the eggplant grew bigger than its Oriental ancestors and acquired the glossy purple color prized by eggplant planters today. The ancient Persians, who ate smoked camel hump, zebra, and Arabian ostrich, were hardly a people to balk at eggplant, and soon devised hundreds of recipes. Latter-day Turks, whose versatile cuisine featured raisins and olives, goats and pigeons, rose water, almonds, yogurt, and powdered pistachio nuts, were said to eat eggplant, cheerfully, at every meal. One dish, eggplant stuffed with pine nuts, was reputedly so overwhelmingly scrumptious that it was known as *Imam Bayildi* ("The Priest Fainted") because such was its gustatory effect.

One dish, eggplant stuffed with pine nuts, was reputedly so overhelmingly scrumptious that it was known as *Imam Bayildi* ("The Priest Fainted") because such was its gustatory effect.

Fainting Turks aside, the introduction of the eggplant to western Europe was a slow and sticky business. From the Middle East, the plant spread across northern Africa to Spain, where, in the twelfth century, a horticulturally accomplished Moorish Spaniard named Ibn-al-awam proudly described four species. It arrived in northern Europe by the sixteenth century, where practically everybody except the dauntless Italians refused to eat it. It did, however, achieve some popularity among dedicated gardeners as an exotic ornamental. Most common early on were small white-fruited varieties—hence the descriptive name "egg" plant—but other types were known. Leon Fuchs (1542) mentions purple and yellow eggplants; Rembert Dodoens (1586), who calls them "unholsome" and claims they filled the body with evil humors, mentions purple and "pale"; Jacques Dalechamp (1587) describes purple, yellow, and "ash-colored" varieties in long, round, and pear shapes. Master herbalist John Gerard mentions white, yellow, and brown eggplants, and spurns them all: "I wish Englishmen to content themselves with meats and sauce of our owne country than with fruit eaten with apparent perill; for doubtless these Raging Apples have a mischevious qualitie, the use whereof is utterly to be foresaken."

The ominous nickname raging, or mad, apple derived from the eggplant's (unwarranted) reputation for inducing instant insanity in the unwary eater. This reputation originated, the story goes, with the first Occidental to sample the new vegetable: tickled with its luscious appearance, he gulped one raw and promptly fell into a fit (the result, one vegetable authority suggests, of acute gastritis). The incident was to haunt the eggplant for centuries. The situation was exacerbated by Sir John Mandeville, an inventive (and possibly invented) fourteenth-century traveller who, along with mermaids, monsters, and considerable imaginary Asian geography, described the Apples of Sodom— delectable-looking purple Levantine fruits that crumbled to ashes when picked. Milton, in *Paradise Lost,* fed these to Lucifer's hapless fallen angels. Today the story survives in *Solanum sodomeum,* a weedy eggplant relative native to the Mediterranean region, commonly known as the Dead Sea apple or the yellow popolo. In addition to insanity, the true eggplant was accused of provoking fever, epilepsy, and lust, and general-purpose avoidance was counselled on the grounds that the fruits dangerously resembled those of the (male) mandrake. None of these less-than-complimentary tales did the European eggplant a bit of good, and culinary acceptance was correspondingly slow.

Among the eggplant's earliest formal appearances on the European table was an Italian banquet thrown by Pope Pius V in 1570. The papal guests started out with marzipan balls, grapes, and prosciutto cooked in wine, then worked their way through spit-roasted skylarks, partridges, pigeons, and boiled calves' feet to a grand finale of quince pastries, pear tarts, cheese, and roasted chestnuts. The eggplant showed up early in the second course, sliced, with quails. During the next century, it continued to edge its way sporadically onto upper-class menus. Its phlegmatic progress may have been due in part to the culinary competition. The seventeenth century was a period of gastronomic discovery, during which the eggplant was outclassed by the introductions of coffee, tea, chocolate, sherbet, turkey, and sparkling champagne.

The ominous nickname raging, or mad, apple derived from the eggplant's (unwarranted) reputation for inducing instant insanity in the unwary eater.

For all this aristocratic company, the Italian eggplant continued to lurk in the shadow of its suspicious past. The modern name *melanzana* comes directly from the older *mala insana,* or mad apple. The plant had better linguistic luck in France, where it was dubbed *aubergine,* a derivation of the ancient Indian *brinjal.* The French acquired the aubergine under the auspices of Louis XIV, who, for all his defects, had spectacular taste in food, mistresses, and gardens. (Aubergine these days has such appealing connotations that the word has been adopted by the fashion industry to indicate an expensive shade of brownish-purple.) The English and Germans adopted the benignly descriptive *eggplant* or *eggfruit* (*Eirfrucht*), and the Jamaicans, who got it from the Spaniards, referred to the fruits cosily as garden eggs. In 1753, Linnaeus grudgingly listed the eggplant as edible, and assigned it for posterity the name *Solanum insanum*—later revised to *Solanum melongena,* which means "soothing mad apple," a nice piece of scientific fence-sitting if there ever was one.

It's not known when *S. melongena* made it to the American side of the Atlantic. Some sources credit Thomas Jefferson, who grew them along with everything else he could lay his hands on in the vast vegetable garden at Monticello. Others hypothesize that it arrived earlier, in the slave ships from western Africa, and was first established along the southern coast, where it was known familiarly as Guinea squash. A final faction, mostly wishfully, claims eggplant was introduced to American diners by Delmonico's, the magnificent New York restaurant founded in 1831 by the Swiss Delmonico brothers, Peter and John. Delmonico's was coupled with the Yosemite Valley by a London newspaper of the 1880s as one of "the two most remarkable bits of scenery in the States." An endless succession of the rich and famous ate there, including Jenny Lind, Louis Napoleon, Abraham Lincoln, Charles Dickens—who routinely drank two bottles of champagne and a glass of brandy for lunch—and Samuel F. B. Morse, who dramatically sent the first telegram from his table. (A reply came back in forty minutes.) Delmonico's also is said to have served up the first American avocados, watercress, and truffles, to say nothing of lobster Newburg and chicken à la king.

Eggplant, by the nineteenth century, was appearing regularly in cookbooks, baked, stuffed, boiled, fried in butter, or, more exotically, stewed in wine and pepper, or pickled in honey and vinegar. Miss Eliza Leslie in her *Directions for Cookery,* following a recipe for eggplant stuffed with chopped tomatoes, announced surprisingly, "Egg-plant is sometimes eaten at dinner, but generally at breakfast." The eggplant breakfast was a predominately southern meal, since the eggplants themselves, like sweet potatoes, peanuts, and okra, were rare in the chilly North. Matters improved with the expansion of the railroads. By 1888, Juliet Corson, author of the now-inconceivable *Family Living on $500* *a Year,* mentioned that city-dwellers could easily get fresh off-season tomatoes, radishes, cucumbers, spinach, Valencia onions, and eggplants.

Eliza Leslie's eggplants were doubtless fried in the very best butter—Miss Leslie came down hard on cooks who reserved their sour or spoiled butter for cooking purposes—and even more certainly soaked up a great deal of that butter during the cooking process. A notable characteristic of the eggplant is its spongy texture. Like a sponge, eggplant tissue contains a large number of intercellular air pockets capable of holding a hefty amount of liquid. Unlike a sponge, however, eggplant reaches a point in the cooking process where it becomes self-squeezing: eventually the heat generated by frying causes the cellular structure to collapse, flattening the air pockets, and extruding the accumulated oil. Popular today for fried eggplant feasts is olive oil, but Middle Eastern gourmets traditionally used *alya,* an oil rendered from the overgrown tails of a special breed of fat-tailed sheep. (The valuable tails were sometimes supported on small two-wheeled wooden carts to protect them from the ordinary wear and tear of sheep life.) Oil-less eggplant possesses a mere fifty calories per cup; sautéed, it's disaster.

Miss Eliza Leslie in her *Directions for Cookery,* following a recipe for eggplant stuffed with chopped tomatoes, announced surprisingly, "Egg-plant is sometimes eaten at dinner, but generally at breakfast."

By the late nineteenth century, the eggplant was a firmly established denizen of the vegetable garden. Vilmorin-Andrieux lists fifteen kinds, ten purple, two white, two green, and one striped in purple and white "lengthways." The famous purple results from a group of ring-shaped chemical compounds called anthocyanins, from the Greek for blue flower. Anthocyanins create the various reds, blues, and purples of blueberries, grapes, red cabbage, and radishes, as well as the unparalleled hue of eggplants. They are both water-soluble and heat-sensitive, which bodes ill for their survival in the cooking pot and explains why boiled red cabbage and strawberry jam often go an unappetizing pinkish-brown.

The loss of purple is seldom a consideration in eggplants, which are often peeled before cooking. Once peeled, however, eggplants, like apples, avocados, bananas, raw potatoes, and pears, develop browning problems. In these fruits, cell disruption by teeth or paring knife releases an enzyme called polyphenoloxidase, which reacts with phenolic compounds in the fruit tissue to form unappetizing brown polymers. In some cases the browning reaction can be slowed down by chilling, though in others cold only makes matters worse; bananas, for example, natives of the balmy tropics, undergo rapid cell damage in the cold and release quantities of polyphenoloxidase, which is why banana peels turn black in the refrigerator. The browning enzyme is also inhibited by the chloride ions in salt, which is useful in the case of potatoes and eggplants, frequently served salted, but less so in the case of apples and pears. Polyphenoloxidase is most effectively blocked, by acidic conditions, which is why a dash of citric acid-rich lemon juice does wonders for preserving the green in guacamole. Browning is also inhibited by ascorbic acid, better known as vitamin C.

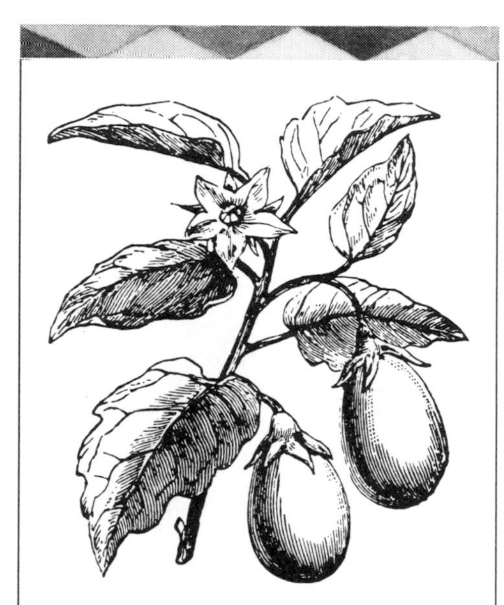

Once peeled, eggplants, like apples, avocados, bananas, raw potatoes, and pears, develop browning problems.

Perhaps the most familiar of eggplant shapes today is the classic bulbous pear, but also available are long cucumber-like and round tomato-like cultivars, small finger-shaped Oriental varieties, and slim foot-long curved varieties nicknamed snake eggplants. The fruits, botanically giant berries, develop from lavender flowers that sprout, unusually, smack out of the plant stem rather than from the leaf axil. The stems may still carry a few scattered spines, remnants of the days when the ancestral eggplant was armed with off-putting inch-long thorns. The fat pears possess green calyces and predominate on vegetable counters; the skinny Oriental types have purple calyces and, though curiously less popular, are tastier. The superiority of the flesh of the Oriental eggplant derives from its slow-developing seeds, which give the Oriental cultivars less of a tendency toward bitterness than their seedier American equivalents. Seediness, incidentally, is the root of all evil in over-the-hill eggplants: those past their prime, identifiable by their brownish tinge and telltale puckery look, are heavily seeded and inedibly bitter. The white eggplants, in the opinion of connoisseurs, should be demoted to ornamental status: though attractive to look at, they're insipid to eat.

Still, insipid is as insipid does, and one of the prime consumers of eggplants is not all that picky. *Leptinotersa decamlineata,* the Colorado potato beetle, is said to like eggplants even better than potatoes. The potato beetle was first described in 1823 by Thomas Say, author of the much-respected *American Entomology.* Say served as chief entomologist at the Philadelphia Museum of Science, where, between bug-collecting expeditions, it was his custom to sleep under the skeleton of a horse in the museum collection. At the time of Say's description, the beetle, yet to discover the delights of potatoes and eggplants, was a fairly innocuous creature living off buffalo bur, a solanaceous weed of the Colorado River basin. It came across potatoes sometime in the mid-nineteenth century, when western settlers started planting them, and moved ravenously eastward in search of them, encountering along the way such additional goodies as eggplants, peppers, and tomatoes. Steadily eating, it reached Illinois in 1864, Ohio in 1869, the East Coast in 1874, and had managed to cross the Atlantic to Europe in 1876.

Adult potato beetles are about half an inch long, striped lengthwise in black and yellow. They're relatively lethargic creatures, enough so to qualify for the first set of one British gardener's admirable anti-pest instructions: "If it moves slowly enough, step on it; if it doesn't, leave it—it'll probably kill something else." The time-honored method for small-scale disposal of potato beetles is to pick them off the plants one by one and drop them into a coffee can of kerosene. Though effective, this method is impractical commercially, for which reason the arsenic-based chemical pesticide Paris green was developed in 1865. Paris green slaughtered potato beetles for about thirty years, until it was replaced by the even more potent lead arsenate in 1892. Both compounds were necessarily used in immense quantities—ten to one hundred pounds per acre—and possessed dire side effects, killing off, along with the beetles, birds, household pets, and an occasional human being. These days the latest in anti-beetle preparations is a "biological pesticide," a new strain of a toxin-producing microorganism, the soil bacterium *Bacillus thuringiensis.* The toxin, a digestive-tract poison, is highly specific, affecting only the target insect species and sparing dogs, children, and other living things. With luck, it may eventually replace the present chemical pesticide of choice, carbaryl, a synthetic organic that, though choosier than arsenic, still does in a number of useful insects, including the valuable honeybee.

If unmolested by arsenic, carbaryl, *B. thuringiensis,* or the kerosene can, female potato bugs go on to mate and lay multiple batches of tiny bright yellow-orange eggs—up to twenty-five hundred a season—which hatch in a week or so to release voracious orange larvae. After two or three weeks of intensive feeding off potato, eggplant, tomato, or pepper plant leaves, the replete larvae dig their way into the soil and pupate, lying in wait for next year's garden. They emerge as hungry adults in the spring.

Perhaps the best literary tribute to the eggplant—of which, admittedly, there aren't many—is Erica Jong's "The Eggplant Epithalamion," seventy lines of eggplant verse in which the subject is referred to aphrodisiacally as "love's dark purple boat."

"Every animal is sad," says Jong, "after eggplant."

PEAS

King John of England, the uncongenial monarch under whom Robin Hood wreaked so much havoc on the rich, died on October 19, 1216. According to the encyclopedia, death was due to dysentery and fever; according to food historians, it was due to overindulgence in peas, seven bowlsful at a single sitting. (Alternatively, the fatal dish was lampreys, unripe peaches, or toad's blood in the royal ale.) If peas, King John could have done better for himself in the way of last meals. The thirteenth-century pea was tough and starchy, much less palatable than the sweet, tender varieties grown today. The earliest peas seem to have been even worse: archaeologists suggest that in order to choke them down, cave-dwellers probably roasted and peeled the seeds like chestnuts.

The pea is such an ancient food plant that its center of origin is uncertain. Most botanists propose a vaguely sizeable area extending from the Mediterranean through the Near East to central Asia. Annoyingly, the oldest pea find to date was made outside of the designated area, at the Spirit Cave site on the Burma-Thailand border; the retrieved peas were radiocarbon-dated to 9750 B.C. Ancient pea remains have been recovered from Swiss lake dwellings and from Neolithic farming villages scattered across Europe, and leftovers from Near Eastern pea feasts have been dated to 7000 B.C. An alternative, but almost certainly inaccurate, proposal locates the primeval pea in China. Chinese legend holds that peas were discovered five thousand years ago by the Emperor Shên Nung, the Chinese "Father of Agriculture." Shên Nung, the story goes, was given to wandering about the Chinese countryside collecting promising food plants, which were then sequentially tested on the palace dogs and the palace servants, and finally sampled by the Emperor himself. Wheat and rice were discovered in this fashion, as well.

The northern Europeans, who didn't acquire peas until the Bronze Age and so scientifically speaking aren't even in the running, have also put in a claim for the original pea. Peas, in Norse legend, arrived on earth as a punishment sent by the god Thor, who, in a fit of pique, dispatched a fleet of dragons with peas in their talons to fill up the wells of his unsatisfactory worshippers. Some of the peas missed and fell on the ground, where they developed into pea plants. The new vegetable was placatingly dedicated to Thor and thereafter eaten only on his day, Thursday. (From then on, when Thor was annoyed, he sent his dwarves to pick the pea vines clean.)

The leguminous plants are distinguished by their ability to fix nitrogen, through symbiotiic *Rhizobia* bacteria located in nodules on their roots. These bacteria are able to convert atmospheric nitrogen to ammonia, which the plant then uses in the synthesis of essential nutrients.

In whatever location, the line between the wild and the domesticated pea is distressingly fine. The cultivated cereal grains, a delight for archaeologists, are distinguished from their wild counterparts by the possession of a nonbrittle rachis—a nonshattering ear—such that the nutritious seeds cannot be disseminated without the helpful hand of man. *Pisum sativum,* the cultivated pea, however, has no comparably definitive feature. Instead, the evolution of the smooth-seeded early cultivated pea from its rough-seeded ancestor was a gradual process, generating a confusing number of questionable intermediate forms. The identity of the ancestral wild pea is unknown and it may be long extinct. However, in 1973, botanical researchers Daniel Zohary and Maria Hopf proposed the wild Near Eastern pea, *Pisum humile,* as the possible ancestral form, on grounds of chromosomal similarity.

Peas, both wild and tame, are legumes, members of the family Leguminosae, which bear their fleshy proteinaceous seeds in a protective pod. Pea relatives include lentils, broad beans, chick peas, soybeans, peanuts, lima beans, kidney beans, carob, and licorice—and, less edibly, clover, wisteria, mimosa, rosewood, and indigo. The leguminous plants are distinguished by their ability to fix nitrogen, through symbiotic *Rhizobia* bacteria located in nodules on their roots. These bacteria are able to convert atmospheric nitrogen to ammonia, which the plant then uses in the synthesis of essential nutrients. Plants without such bacterial buddies require another nitrogen source, often, in the case of modern commercial crops, provided by expensive chemical fertilizers. A time-honored alternative to such unnatural additives is crop rotation; since Roman times, farmers have periodically planted their fields with *Rhizobia*-toting legumes to add fixed nitrogen to the soil. In terms of nitrogen-fixing ability, not all *Rhizobia* strains are created equal, and researchers have recently discovered that *R. japonicum,* which inhabits soybean roots, is a champion of its kind. In general during nitrogen fixation, about 25 percent of the energy needed to fuel the process is lost as a waste product in the form of hydrogen gas. *R. japonicum,* however, has an efficient hydrogen recycling system, controlled by a battery of *hup*—for *h*ydrogen *up*take—genes, which allows it to recapture much of this lost energy. More energy is thus available to the soybean plant, which pumps it into leaves and protein-packed seeds. The soybean, a legendarily nutritious legume, contains up to 38 percent protein. Less spectacular beans and peas have protein concentrations of 17–25 percent—still about twice as much as most cereals and enough to explain the legume's traditional nickname, the poor man's meat.

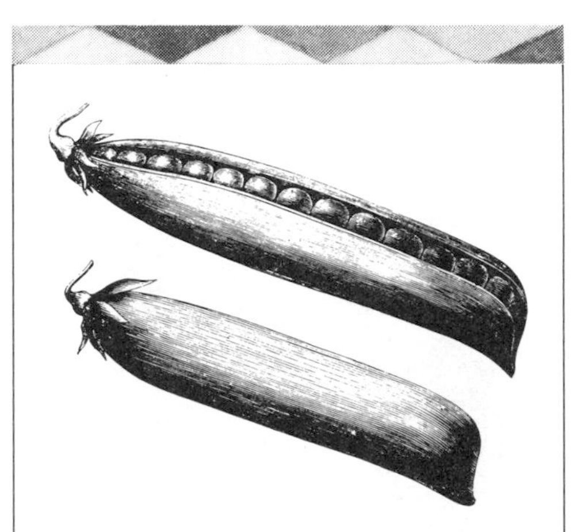

Hot pea soup was peddled in the streets of Athens; fried peas were sold to spectators, in lieu of popcorn, at the Roman circus and in theaters.

The Greeks and Romans grew peas. Hot pea soup was peddled in the streets of Athens; fried peas were sold to spectators, in lieu of popcorn, at the Roman circus and in theaters. Upper-class Romans ate their peas with salted whale meat; the lower classes ate theirs in porridge. Most peas in ancient times were consumed dried, the drying process being considered essential to cure the pea of its "noxious and stomach-destroying" qualities. Uncured peas were occasionally left on the vines by farmers, with the intention of poisoning pestiferous rabbits, who thus may have gotten the most out of the classical pea. Dried peas remained the rule for the next several centuries, convenient because peas thus treated could be stored almost indefinitely, for winter use, as ships' stores, or as a bulwark against famine. Dried peas were used to piece out wheat flour, or were boiled to make the ubiquitous pease porridge that was eaten hot, cold, and in the pot nine days old. Not always a simple dish, one recipe of the eighteenth century began with beef broth in which was boiled a chunk of bacon and a sheep's head, then called for unmeasured quantities of nutmeg, cloves, ginger root, pepper, mint, marjoram, thyme, leeks, spinach, lettuce, beets, onions, old Cheshire Cheese (grated), "sallery," turnips, and "a good quantity" of peas. To obtain a "high taste," the cook recommended tossing an old pigeon in with the bacon. A more conventional "Peas-Porridge" is described by Hannah Glasse in her modestly titled "The Art of Cooking Made Plain and Easy, Which far Exceeds any thing of the Kind ever yet Published" (1947):

Take a Quart of Green Peas, put to them a Quart of Water, a Bunch of dry'd Mint, and a little Salt. Let them boil till the Peas are quite Tender, then put in some beaten Pepper, a Piece of Butter as big as a Wallnut, rolled in Flour; stir it all together, and let it boil a few Minutes. Then add two Quarts of Milk, let it boil a quarter of an Hour, take out the Mint, and serve it up.

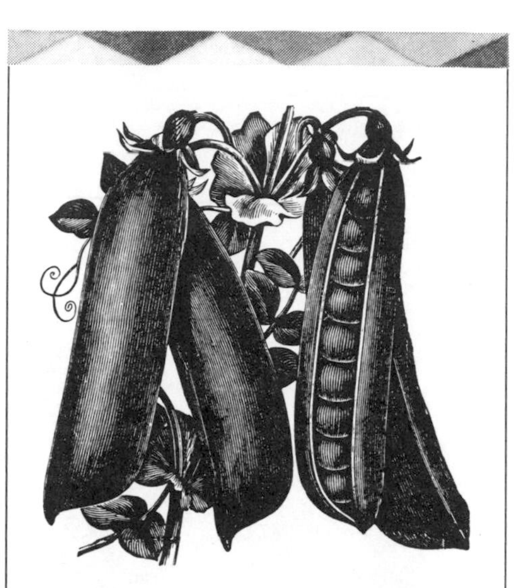

A May 1695 letter of Madame de Maintenon, last and most successful of Louis XIV's mistresses (he married her) reads: "The subject of peas is being treated at great length: impatience to eat them, the pleasure of having eaten them, and the longing to eat them again are the three points about which our princes have been talking for four days."

The cooking of peas with mint may have originated to disguise the starchy taste of the early smooth-seeded peas; the custom, however, persisted into the nineteenth century, well after the introduction of sweeter, wrinkle-seeded pea varieties. In America, the cooking style may have been encouraged by the prevalance of mint, so common in gardens that an anti-julep temperance movement of the 1830s campaigned under the slogan "Every bed of mint must be uprooted!" By the end of the century, progressive cooks were advising preparation of peas in the "American mode"—boiling in plain water—and Mary Henderson in her *Practical Dinner Giving* (1882), wrote damningly, "This cooking of pease with mint is a good way of utterly destroying the delicious natural flavor of the pea."

The peas that eliminated the necessity for these concealing doses of mint appeared in the sixteenth century, a high-sugar mutant variety known as Rouncivals after the Roncevaux in the French Pyrenees, where Roland trounced the Saracens. The new sweet-tasting pea was tailormade for the new custom of eating peas fresh—and, in fact, the fresher the better, since like corn, peas deteriorate rapidly after picking. The modern pea, 25 percent sucrose by weight, loses nearly half of its sugar in six hours at room temperature. The new pea soon reached England, where it was listed by John Parkinson among the nine pea varieties described in *Paradisi in sole* (1629): "the Runcival, the Green Hasting, the Sugar Pease, Spotted Pease, Gray Pease, White Hasting, the Pease with skins, the Scottish or Tufted Pease, sometimes called the Rose Peas, the early or French Pease, called the Fulham Peas." By the seventeenth century, green-pea-eating had become a positive passion. A May 1695 letter of Madame de Maintenon, last and most successful of Louis XIV's mistresses (he married her) reads: "The subject of peas is being treated at great length: impatience to eat them, the pleasure of having eaten them, and the longing to eat them again are the three points about which our princes have been talking for four days. There are some ladies who, after having supped with the king, and well supped too, help themselves to peas at home before going to bed at the risk of indigestion. It is a fad, a fury." The peas were dunked, pod and all, into a dish of sauce, and then eaten out of the shell.

The first deliberate attempts at pea breeding employed these delectable wrinkle-seeded peas. Performed by Englishman Thomas Knight, these artificial crosses resulted in an improvement on the Rouncival called the Marrow pea, which was introduced in the 1780s and became rapidly popular. By the nineteenth century, there were dozens of different garden peas available. William Cobbett, peevish author of *The English Gardener* (1833) remarks, "As to sorts of peas, the earliest is the early-frame, then comes the early-charlton, then the blue-prussian and the hotspur, then the dwarf-marrowfat, and the tall-marrowfat, then the knight's pea." (Cobbett's "knight" refers to the gardening Thomas.) He concludes, somewhat snappily, "There are several others, but here are quite enough for any garden in the world." By the end of the century, matters were even worse for the horticulturally indecisive. Vilmorin-Andrieux's *The Vegetable Garden* describes 170 different varieties of peas, classed as either shelling peas (subclasses: round and wrinkled, tall, half-dwarf, and dwarf) or sugar peas. Peas appeared with increasingly superlative titles, of the sort reserved today for hamburgers and used cars: Wonder, Marvel, Ne Plus Ultra, Pride of the Market, Best of All. Breeder Thomas Laxton—

whose peas are still around today—continually overtook himself, creating the Alpha, the Omega, and the Supplanter.

A snow pea past its prime tends to twist arthritically due to the lack of supportive parchment—the "bones" of the pod—and concomitantly develops an unpleasant taste.

The pea arrived in North America before reaching these glorious heights. It was first planted by Columbus, in his 1493 garden on Isabella Island, and first reached New England in 1602 when Captain Bartholomew Gosnold put in a few rows on the Maine island of Cuttyhunk. The first colonists came well equipped with peas. The Pilgrim peas failed the first year (so did the barley, optimistically intended for English beer), but by 1629 the governor's garden at Massachusetts Bay, according to the Reverend Francis Higginson, was growing green peas "as good as I ever eat in England." John Smith gloated over the pea crop at Jamestown ("Pease dry everywhere"), and peas figured routinely in the lists of supplies recommended for newcomers by seasoned settlers. One such, dated 1635, calls for "three paire of Stockings, six paire of Shooes, one gallon of Aquavitae, one bushell of Pease."

The traditional *pease* lost its plural in the New World. Evolving from the Latin *pisum,* the word the early English used was *peason* (also insouciantly used to mean beans), eventually shortened to *pease.* In the colonies, the word at some point was further cut to the singular *pea.* It was still *pease,* however, in 1770, when Benjamin Franklin, who kept up a correspondence with the famed Pennsylvania plantsman John Bartram, sent him "some green dry Pease, highly esteemed here as the best for making pease soup," some rhubarb seed, assorted turnips, and, mysteriously, "some Chinese Garavances, with Father Navaretta's account of the universal use of a cheese made from them, in China." Garavance is a Spanish word, usually used for the chick-pea (*Cicer arietinum*), a legume essentially impossible to turn into any sort of cheese. One suggestion, by agricultural historian U. P. Hedrick, is that Franklin's "cheese" was tofu, and that Franklin may have introduced the soybean to America. Thomas Jefferson, who also grew Garavances, planted thirty different kinds of peas at Monticello. Peas, Jefferson claimed, were his favorite garden vegetable, and during his presidential term in Washington, he wrote a touching reminder home to his Virginia overseer to plant the "Ravenscroft peas, which you will find in a canister in my closet." He also planted Charlton, Rouncival, and Dutch Admiral peas, "bush" and "bunch" peas, and hog and cow peas, the last also called black-eyed peas, which are really beans.

Shelling peas, both smooth- and wrinkle-seeded, develop within inedible pods, rendered unchewably indigestible by their fibrous parchment lining. Parchmentless edible-podded peas of the sort commonly known as Chinese or snow peas seem to have been developed by the Dutch. The earliest mention dates to 1536. It is likely that these are the peas, expensively imported from Holland, that were considered such a treat at the court of Elizabeth I, and that were eaten as *mange-tout,* meaning "eat all" in sixteenth- and seventeenth-century France. The classic snow pea, however, must be picked on cue, at what breeders call the "slab-pod" stage, before the inner peas begin to bulge out and stringiness develops. A snow pea past its prime tends to twist arthritically due to the lack of supportive parchment—the "bones" of the pod—and concomitantly develops an unpleasant taste.

These problems have been eliminated for modern gardeners by the advent of the Sugar Snap pea, a hybrid developed by breeder Calvin Lamborn of the Gallatin Valley Seed Company in Twin Falls, Idaho. Lamborn's pea is the result of a cross between a tough-podded mutant of a processing pea called Dark Skinned Perfection and a conventional snow pea. His original intent was to solve the snow pea twisting problem by adding genetic material from the strong-prodded mutant strain; the unexpected outcome, a tasty sugar pea, juicily edible into full maturity, is now touted as a triumph for pea breeding. The "snap" designation comes from the pea's breaking characteristics: it cracks neatly in two, like a green bean. Since the introduction of Sugar Snap in 1979, Lamborn has further perfected his vegetable brainchild, developing disease-resistant and dwarf versions of the parent pea—the original vines grew upwards of six feet long—and producing Sugar Snaps without the tough dorsal and ventral strings. The most successful of the stringless Sugar Snaps—SugarBon, SugarMel, and SugarRac—are named for

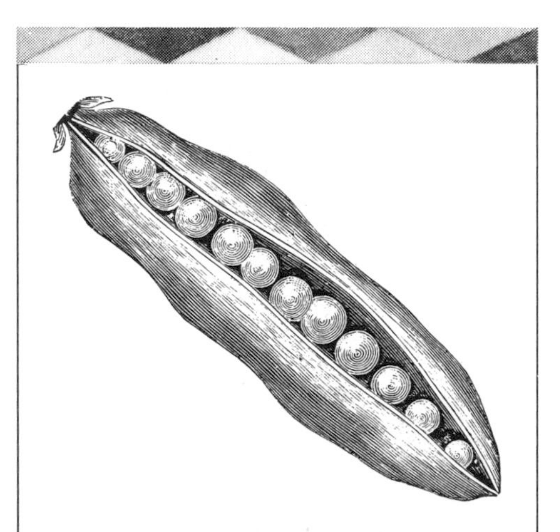

Since the introduction of Sugar Snap in 1979, Lamborn has further perfected his vegetable brainchild, developing disease-resistant and dwarf versions of the parent pea—the original vines grew upward of six feet long—and producing Sugar Snaps without the tough dorsal and ventral strings.

Lamborn's wife and daughters.

Sugar Snap peas, though best fresh off the vine, are also suitable for freezing, a fate that has overtaken 90 percent of the national pea crop since Clarence Birdseye came up with his commercial freezing procedure in 1929. These days, peas, frozen and fresh, constitute the second largest vegetable crop in the United States (after corn), and the twenty-eighth largest in the world. They appear, barely, on the Food and Agriculture Organization (FAO) list of the world's major crops, nearly at the bottom, between beans and mangoes.

If plant scientists have their way, however, peas—or at least choice genetic bits of peas—may soon have a larger part to play. Australian scientists at the CSIRO Division of Plant Industry are presently attempting to isolate the genes for one of the principal pea proteins, p-albumin-1, and to insert them into alfalfa and other forage plants for sheep. The goal of this genetic tinkering is more and better wool. The wool fiber is heavy in cysteine, a sulfur-containing amino acid. Therefore, to produce wool, the sheep must have a diet of cysteine-rich foods. The sheep, however, like the cow and the camel, is a ruminant, possessed of multiple stomachs that allow it to digest humanly uncongenial leaves and grasses. Converting this greenery to absorbable form is a heavy-duty process, and often, in the first stomach (the rumen), the sulfur amino acids are degraded and lost. P-albumin-1 contains 12 percent sulfur amino acids (11 percent cysteine and 1 percent methionine) and is unusual in that it resists destruction in the rumen. If this pea protein makes up 3–4 percent of the total alfalfa leaf protein, scientists hypothesize, sheep fed on such plants could exhibit a 50 percent increase in wool production. To date, Australia is still a good way from waving fields of pea-protein-supplemented alfalfa; plant geneticists are presently attempting to insert the relative pea genes into cells from high-quality alfalfa plants and to screen the progeny for production of p-albumin-1. One worry for the future: high-wool sheep fed on sulfur-rich alfalfa may run into trouble in the rain. All that wool, researchers predict, will be heavy when wet, and superwoolly sheep may have a tendency to bog down.

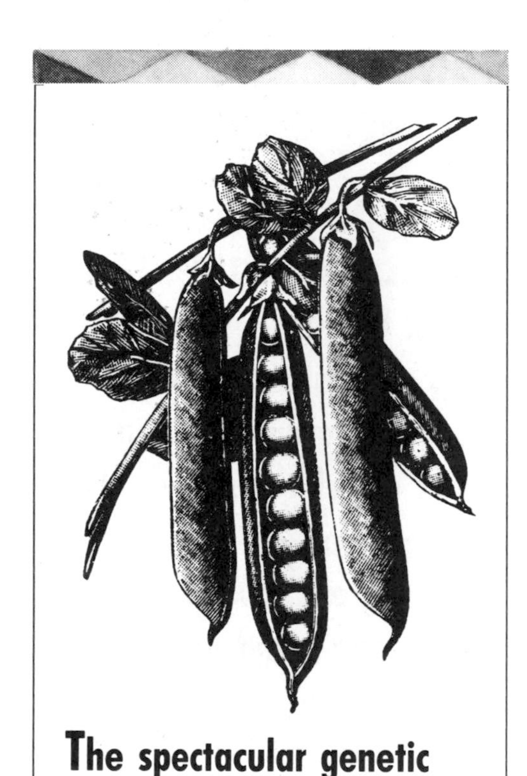

The spectacular genetic advances of the present day are all, in the final analysis, in debt to the garden pea.

The spectacular genetic advances of the present day are all, in the final analysis, in debt to the garden pea. The modern scicnce of genetics is founded in large part on the work of Gregor Mendel, an Austrian monk, who began his landmark experiments in the monastery garden in Brünn in 1856. Mendel's pea plant crosses, based on simple defined characteristics such as seed shape (round or wrinkled), stem length (tall or short), and color (green or yellow), led to a number of fundamental conclusions about the behavior of heritable elements—which would not, for another one hundred years, come to be known as *genes*. From Mendel's peas came our modern concepts of dominant and recessive, phenotype and genotype. For example, the color green in peas, Mendel found, was dominant over yellow; hence an initial cross between a purebred green and a purebred yellow plant will yield only green offspring. All the offspring will have the same genotype; each will have received one gene from the dominant green parent (GG) and one from the recessive yellow parent (gg). Thus all will be heterozygotes (Gg). When these members of the first filial (Fl) generation are crossed in their turn, the genetic elements separate and re-assort to yield both green and yellow progeny in a ratio of 3:1. One of these will be identical to the original green parent (GG); two, though possessing the same phenotype (they *look* green), will havc a different genotype (Gg); and one will be yellow (gg). From such results, accumulated over years of painstaking hand-pollinization, Mendel evolved his famous Laws of Segregation and Independent Assortment. In the off season, he spent his time teaching natural science at the Brünn high school, though he never managed to pass the examinations to obtain a teaching certificate.

Finally, there is the meteorological pea, which figures in the pea-souper, a dense yellowish fog impossible to see through, in the face of which one wears a pea jacket. Gardeners in such weather tend to stay indoors, slothfully reading the seed catalogs.

BEANS

In the late nineteenth century, an enterprising American distributor marketed his green beans— then coming into vegetable vogue—as "the Ninth Wonder of the World." It's not certain how this appealing slogan affected the case-hardened American consumer, but it did catch the jealous eye of P. T. Barnum, who claimed that he had coined the term himself, to describe such phenomena as General Tom Thumb, Chang and Eng the Siamese Twins, the Fejee Mermaid, and Jumbo the Elephant. Barnum sued and beat the bean, which subsequently qualified its Wonder status by adding a cautious "of Food" in smaller print.

The wondrous bean that figured in this legal brouhaha was *Phaseolus vulgaris,* the so-called American, French, kidney, or common green bean. *P. vulgaris* has been cultivated for thousands of years, perhaps originally domesticated from a wild ancestor resembling the uncivilized-sounding *P. aborigineus,* now native to Argentina and Brazil. Bean seeds from archaeological sites in Peru and Mexico have been radiocarbon-dated respectively to 8000 and 5000 B.C., and by the time of the Spanish conquest, Montezuma was taking in five thousand tons of cultivated beans a year in tribute from his devoted subjects. Columbus noticed the American beans ("very different from those of Spain") in Cuba on his first voyage to the New World. Giovanni da Verrazano, flushed with success after his North American voyage of 1524, described the Indian beans as "of good and pleasant taste"; Samuel de Champlain, in his 1605 account of the Indians of the Kennebec region in Maine, mentioned their cultivation of "Brazilian beans" in different colors, three or four of which were planted in each hill of corn to grow up the supporting cornstalk. John Hariot mentioned beans (in two sizes) grown by the Indians on Roanoke Island, and John Smith and Powhatan went on record as sharing a macho meal of beans and brandy in the early days of the settlement at Jamestown.

Varieties are found in a spectacular array of yellows, tans, greens, pinks, maroons, and purples, often mottled, spotted, streaked, or splashed in contrasting shades.

It seems to have been the Spanish who brought the American bean to Europe, sometime in the sixteenth century, where it was initially grown as ornamental. At some point, one improbable-sounding story goes, a bunch of the ornamental pods accidentally toppled into a soup pot, where they remained long enough to be cooked and eaten, thus introducing the American bean to European cuisine. In France, beans became an integral part of a traditional meat-and-vegetable stew known as *hericoq,* so much so that they eventually took over the name for their own, which is where we get *haricot,* or French bean. (An alternative source says *haricot* comes from the Aztec word for bean, *ayacotl.*) To distinguish the haricot dried from the haricot fresh, precise cooks adopted the term *haricot vert* ("green bean"). The *haricot vert* and associated recipes arrived in England in company with a group of fleeing French Huguenots during the reign of Elizabeth I. The Huguenots, in gratitude for religious freedom, made a gift of green beans to the Queen, who found them "much engaging to the royal taste" and ordered some planted in her garden at Hampton Court. A brief movement flared up among patriotic farmers to rename the new vegetable the Elizabeth bean, but the name never quite took, and, in English gardens, the American bean remained French.

Elizabeth's bean, *P. vulgaris,* is the dominant species of the American beans, all of which are members of the nitrogen-fixing Leguminosae family. The proper name *Phaseolus* comes from the Latin for *little boat,* because of the vaguely canoe-like shape of the seed pods; the common name *kidney bean* comes from the anatomically suggestive shape of the seeds themselves. *P. vulgaris* is a bean of almost incredible virtuosity. "The stocke of kindred of the kidney Beane," commented Edward Johnson in 1636, "are wonderfully many." Botanists restrict *P. vulgaris* seeds to five main color groups—white, black, red, ochre, and brown—but, in actual practice, varieties are found in a spectacular array of yellows, tans, greens, pinks, maroons, and purples, often mottled, spotted, streaked, or splashed in contrasting shades. Seeds also come in a range of shapes, including round, oval, flat, and long, as well as kidney-shaped. Pods, flat or curved, are also colored in green, yellow, purple, red, or white, sometimes speckled or splotched.

Perhaps because of this decorative variety—they look so nice in glass jars—beans rate high with heirloom seed collectors. Beans, in fact, have a collectors' society all to themselves, founded in the 1970s by bean-lover John Withee, who got into the business in a fit of annoyance after failing to track down the proper bean—the Yellow Eye—for simmering up the Maine version of baked beans. Withee named his organization Wanigan Associates, after the cooking shacks—wanigans—used by Maine lumberjacks. In 1981, with well over a thousand varieties of beans under his belt, Withee turned Wanigan Associates over to Kent Whealy's Seed Savers' Exchange, a national association of heirloom seed collectors based in the Corn Belt of Iowa. Among Withee's thousand are such old-fashioned favorites as the soldier bean, white with an enlisted-man-like splotch in maroon at the hilum; the Jacob's cattle bean, also called the Coach Dog or Dalmation, white speckled in dark red; the Wren's Egg, which looks like a wren's egg; the cranberry, which is cranberry-colored; the Black Turtle bean, star of black bean soup; the pinto bean, which the cowboys called "Mexican strawberries"; and the Oregon giant bean, grown in the gardens of the western pioneers, which features eight-inch-long pods dotted with pale blue.

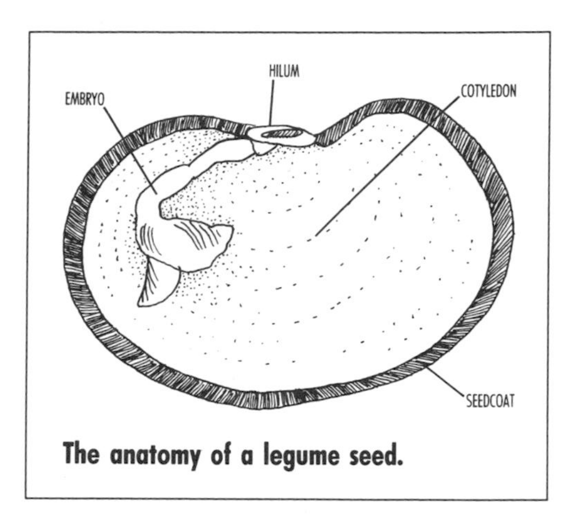

The anatomy of a legume seed.

P. vulgaris had spread throughout North America along the old Indian trails well before the arrival of the Europeans, and had established itself as one of the Indian vegetable triad, an inevitable companion of corn and squash. The Indians generously passed their beans and the means of preparing them on to the settlers. The Cherokees picked beans green, wove them into long chains, and hung them out to dry in the sun, to a consistency that led the Europeans to call them "leather-britches" beans. The New England Indians dried beans on the vine and used them to make the original of

Boston baked beans: The beans were soaked in water until softened, then mixed with bear fat and maple sugar and baked overnight in a "beanhole"—a hole dug in the earth and lined with hot stones. The slow overnight baking particularly appealed to the colonists at Massachusetts Bay, for whom cooking was prohibited on the Sabbath Day. Beans, tossed into the pot on Saturday night, seemed a neat solution to the problem of Sunday dinner, and soon became a Boston tradition. Boston, accordingly, acquired the nickname Beantown. The city's predilection was commemorated in verse by John Collins Bossidy in a high-flown moment at the Holy Cross Alumni dinner of 1910:

And this is good old Boston,
The home of the bean and the cod,
Where the Lowells talk only to the Cabots,
And the Cabots talk only to God.

Some Bostonian bean-lovers even routinely poured the water off their boiled beans, and drank it in cups, as "coffee."

BAKED PORK AND BEANS

From Mrs. Lincoln's Boston Cook Book, *1891*

◆ Soak one quart of pea beans in cold water over night. In the morning put them into fresh cold water, and simmer till soft enough to pierce with a pin, being careful not to let them boil enough to break. If you like, boil one onion with them. When soft, turn them into a colander, and pour cold water through them. Place them with the onion in a beanpot. Pour boiling water over one quarter of a pound of salt pork, part fat and part lean; scrape the rind till white. Cut the rind in half-inch strips; bury the pork in the beans, leaving only the rind exposed. Mix one teaspoonful of salt—more, if the pork is not very salt—and one teaspoonful of mustard with one quarter of a cup of molasses. Fill the cup with hot water, and when well mixed pour it over the beans; add enough more water to cover them. Keep them covered with water until the last hour; then lift the pork to the surface and let it crisp. Bake eight hours in a moderate oven. Use more salt and one third of a cup of butter if you dislike pork, or use half a pound of fat and lean corned beef.

Also popular among colonial bean-eaters was succotash—from the Indian *misickquatash*—originally a mix of kidney beans, maize, and bear grease, but in bearless later-day versions, usually corn and lima beans. Lima beans, second of the four major cultivated *Phaseolus* species, are officially known as *P. lunatus,* a scientific catchall that covers both large-seeded limas and small-seeded sieva or butter beans. Lima beans are named after their city of approximate origin in Peru, where archaeologists estimate they were under cultivation six thousand years ago. The domesticated lima bean thus considerably predates Lima, which was founded in 1535 by Francisco Pizarro on the site of an older Indian village called Rimac, and should more accurately be called the Rimac bean. The Spaniards sent samples of the local beans back to Europe, and then distributed them, as a sideline of their numerous voyages of exploration, to the Philippines, Asia, Brazil, and Africa. Small-seeded *P. lunatus* varieties seem to have developed later and farther north; the earliest finds date to A.D. 800 in Mexico.

The beans are said to have reached the United States in the early nineteenth century, picked up in Peru by a naval officer named John Harris, who first grew them in his garden in Chester, New York. If so, they caught on like wildfire; both lima and sieva beans are mentioned in the 1812 diary of Benjamin Goddard, a gardening resident of Brookline, Massachusetts. Chances are there were multiple introductions; the varieties grown today can be traced back to a scattering of original South American imports. The earliest limas were pole beans; bush limas seem to have appeared by freak chance in the 1870s, when some alert bean collector stumbled upon a dwarf plant along a Virginia roadside.

Most spectacular of the cultivated American beans is certainly the scarlet runner, *P. coccineus,* which was first domesticated in Central America and Mexico. Nicknamed painted-lady from its gaudy flowers, the bean was first adopted by Europeans as an ornamental. It was grown in sixteenth-century England as "garden smilax," so-called because it twines, unusually, from right to left like smilax or honeysuckle. John Gerard grew it in his garden, on decoratively positioned poles, though it was equally popular over the "arbors of banqueting places," to lend a note of glorious color to the upper-echelon picnic. A white-flowered variety, the Dutch case-knife bean, was a favorite in eighteenth-century gardens.

The ancient *Phaseolus* of the Southwest is still known today as *P. acutifolius,* the tepary (or, occasionally, Texas) bean. The tepary is a rapid grower, notably resistant to drought, ideally suited to the hot, dry climates of western Texas, Arizona, and New Mexico. It was domesticated in Mexico by at least 5000 B.C. and was intensively cultivated by the Hopi Indians of the American Southwest, who deliberately selected for the widest possible range of colors: yellow, tawny, brown, garnet, bluc-black, white, and variously speckled. Tepary beans, before the advent of the playing card and the poker chip, figured as counters in an ancient Indian gambling game.

Amelia Simmons in her 1796 *American Cookery*—the longest recipe in which is a three-page account of how to dress a turtle—differentiates among string, shell, and dried beans, the three uses to which the edible bean

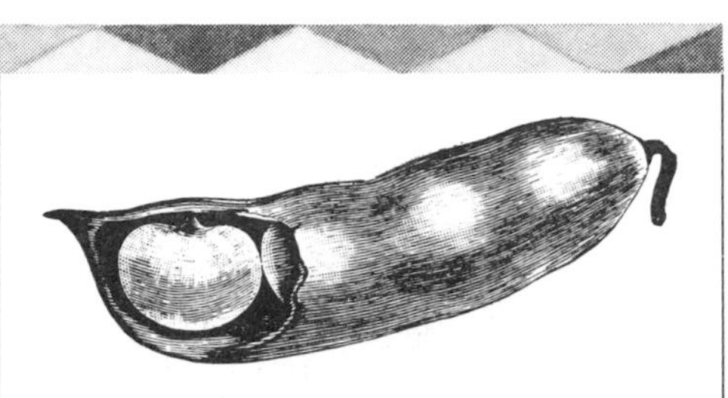

Lima beans are named after their city of approximate origin in Peru, where archaeologists estimate they were under cultivation six thousand years ago. The domesticated lima bean thus considerably predates Lima, which was founded in 1535 by Francisco Pizarro on the site of an older Indian village called Rimac, and should be called the Rimac bean.

is still put today. The string bean—stringless since 1894—is eaten whole, green and immature, while the outer pod is still tender. At this stage, the beans are also called snaps because they crack crisply in two when broken. Not all are actually green: the wax or golden-podded bean was introduced in the 1830s, developed by sequentially selecting for lightened pod color. Most contain a bit less vitamin B than their green relations, but are otherwise equivalent. According to Bert Greene, author of the entrancing cookbook *Greene on Greens* (1984), string beans have been a commercial crop in this country only since 1836, when a farmer from Utica, New York, mortgaged the family home in order to import bean seeds from France. His venture was successful, green beans flooded onto American tables, and ten years later the unforeclosed farmer, now rolling, packed up and relocated, bag and baggage, to the Riviera. Colonial string beans were mostly pole types, often planted in the cornfields Indian-style, so as to clamber up the cornstalks. Bush string beans, though apparently cultivated by the North American Indians (the Omahas, for example, raised "walking beans" and "beans-not walking'), were rare before the nineteenth century, when Grant Thorburn offered one of the first named bush varieties in 1822. It was called Refugee, since it was among the beans brought to England by refuge-seeking French Huguenots in the 1500's.

String beans have been a commercial crop in this country only since 1836, when a farmer from Utica, New York, mortgaged the family home in order to import bean seeds from France. His venture was successful, green beans flooded onto American tables, and ten years later the unforeclosed farmer, now rolling, packed up and relocated, bag and baggage, to the Riviera.

Somewhat older, adolescent beans are eaten as "shell" beans. By the shell stage, nine to eleven weeks after planting, the bean pods have become inedibly rubbery, but the enclosed seeds are still tender and immature, prime candidates for the cooking pot.

The bean reaches adulthood—the dry-bean stage—after twelve to fourteen frost-free weeks on the vine. Designated "best for winter use" by Miss Simmons, these include such traditional baking beans as John Withee's Yellow Eye and the small white oval-shaped bean so commonly used in ships' stores that it is best known as the navy bean. *American Cookery* lists nine varieties of beans, among them the Clabboard, the Crambury ("rich but not universally approved"), the Six Weeks (yellowish and "tolerable"), the Lazy ("tough"), the Windsor, and the English or horse bean, so easily cultivated that it "may be raised by boys" but otherwise had seemingly little to recommend it.

The Windsor and English beans, so cavalierly written off by the first American cookbook, are both varieties of *Vicia faba,* the broad or fava bean. The Windsor (var. *major*) is the broad bean proper, the common eating bean; the horse bean, a smaller version, is raised primarily as animal forage, which likely explains Amelia Simmons' lack of enthusiasm. The fava is an Old World legume, originating in the Near East and domesticated sometime in the late Neolithic period. It appears in Egyptian tombs and in the ruins of Troy, in the Old Testament (Ezekiel ate them between prophecies), and on the tables of the Romans. The Greeks used fava beans as voting tokens during magisterial elections, a custom later remarked upon by Plutarch, whose proverbial dictum "Abstain from beans" meant keep out of politics. Fava beans reached China in the first century A.D., via the ancient Silk Road, and nowadays China is the world's top producer, turning out about 70 percent of the international broad bean crop. The United States, perversely, is now the prime grower of the Asian soybean, and Africa is a major producer of the South American lima bean, which seems to indicate that these things average out.

In the Middle Ages, beans were such an essential article of diet that the penalty for robbing a beanfield was death. In England, ghosts were said to fear broad beans—to banish spectres, one spat beans at them—and those same beans, roasted, were believed efficacious in the treatment of toothache and smallpox. "Three blue beans in a blue bladder" was a tongue-twister long before she sold seashells by the seashore, and lunacy was said to be on the rise when the beans were in blossom. Those foolish enough to sleep all night in a beanfield supposedly woke up irrevocably insane, though history seems a bit weak on medical documentation. Bean-eating hens were said to stop laying eggs. Hippocrates thought beans injurious to the eyesight; Pythagoras, author of the famous triangular theorem so prominent in trigonometry textbooks, forbade the eat-

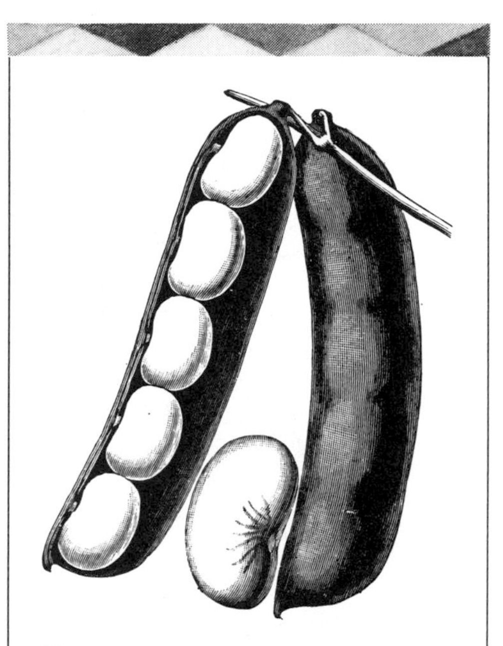

Fava beans reached China in the first century A.D., via the ancient Silk Road, and nowadays China is the world's top producer, turning out about 70 percent of the international broad bean crop. The United States, perversely, is now the prime grower of the Asian soybean, and Africa is a major producer of the South American lima bean.

ing of beans by his followers. The Pythagorean bean ban is said to derive from the belief in the transmigration of souls; human beings could thus be reborn as beans, which made consumption ethically risky.

Some of the fava bean's bad press may be directly or indirectly linked to the disease syndrome known as favism. Particularly common in individuals of Mediterranean ancestry, favism results from a deficiency in the enzyme glucose-6-phosphate dehydrogenase. The deficiency, carried on the X chromosome, renders the red blood cells of unlucky males and homozygous females sensitive to hemolysis by the oxidants found in fava beans. Fava bean consumption, or even simply a stroll through the field when the bean plants are in flower, thus brings on a severe allergic-type reaction, and in extreme cases rapid hemolytic anemia, shock, and possible death. Fava beans also contain a considerable concentration (0.25 percent by weight) of L-DOPA, a chemical neurotransmitter used today as a pharmaceutical in the treatment of Parkinson's disease. As such, favas are the principal commercial source of L-DOPA, and the presence of the compound may account for the classical reports of sleep disturbances, heavy dreaming, and enhanced sexuality associated with bean-eating.

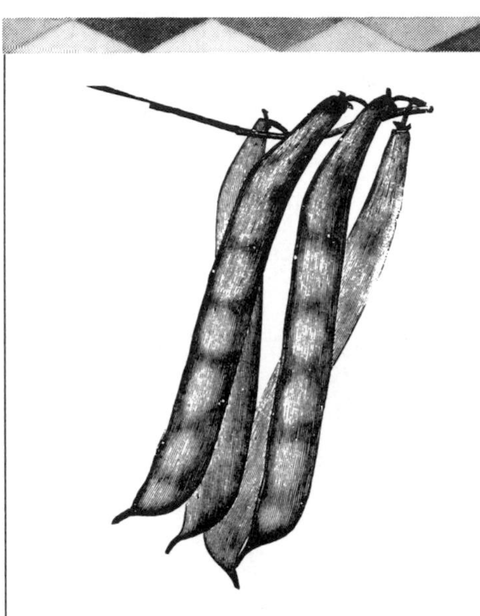

Both Old and New World beans—and, to be fair, bran, onions, cucumbers, raisins, cauliflower, lettuce, coffee, and dark beer—have a reputation for eliciting a condition known delicately in the sixteenth century as "windinesse."

In a less medically exalted level, both Old and New World beans—and, to be fair, bran, onions, cucumbers, raisins, cauliflower, lettuce, coffee, and dark beer—have a reputation for eliciting a condition known delicately in the sixteenth century as "windinesse." Flatulence appears to have been a pressing social concern of the time; Sir Robert Burton in his 1621 *Anatomy of Melancholy* lists sixty-four proposed remedies for sufferers. Nowadays beans continue to do their worst. Their embarrassing aftereffects are now known to be due to an assortment of oligosaccharides—short chains of two to ten linked sugars—which the body is unable to break down into metabolizable form. The bacteria of the lower intestinal tract, however, can digest these tidbits just fine, producing in the process an accumulation of bloating gas. (Beans for this reason are particularly off-limits to astronauts.) A possible solution to the bean problem has been recently discovered by researchers at the Bhaba Atomic Research Centre in Bombay, India. The test subject was *Phaseolus areus,* a bean commonly known as green gram, whose oligosaccharide content consists of 34 percent stachyose, 15 percent raffinose, and 10 percent verbacose, plus 40 percent sucrose, a manageable disaccharide cleaved into its digestible component sugars by the enzyme sucrase found in the human small intestine. Scientists V. S. Rao and V. K. Vakil irradiated green gram seeds with gamma rays from a cobalt-60 source; then analyzed the zapped beans for sugar content. They found that the gamma treatment had little effect on sucrose, but reduced the verbacose concentration by 18 percent and cut concentrations of stachyose and raffinose, the worst offenders, in half. Radiation exerts its deflating effect by weakening the bonds between the linked sugar molecules, making them more susceptible to enzyme cleavage.

While hardly the pinnacle of social acceptability, flatulence is not ordinarily dangerous—except perhaps in the isolated case of a Roman aristocrat under the Emperor Claudius who reportedly endangered his health by embarrassed retention. Other bean components may be less benign. Foremost among the evils are the cyanogens, harmless sugar complexes that in the presence of a specific enzyme are cleaved to release cyanide, an effective and deadly inhibitor of the respiratory system. Cyanogens are found in the seeds of apples, pears, peaches, and apricots, as well as in lima and kidney beans—which last pair owe their appealing flavor to a soupçon of cyanide. Wild relatives generally are higher in cyanogens than the cultivated bean, and some cultivated varieties have more poisonous potential than others. The colored lima beans, for example, of the sort popular in early Peru, contain up to thirty times the cyanogen concentration of the all-white lima beans grown in this country. Along with the cyanogens, bean seeds contain protease inhibitors— complex protein molecules that interfere with the enzymatic processes of digestion—and lectins, which bind to sugar receptors on the surfaces of intestinal cells, with ensuing ill effects. Black Turtle beans, for example, contain a hefty dose of the toxic lectin phytohemagglutinin, which induces a lethal clotting of the blood. Phytohemagglutinin luckily is defused by cooking, which takes black bean soup off the hook. Perhaps, as one anthropological theory suggests, cooking developed in the first place to detoxify the otherwise irresistibly nutritious seeds of wild legumes.

Despite their assorted negative qualities, beans are highly nutritious, containing up to 22 percent protein, plus miscellaneous minerals and vitamin A. In comparison to the egg—the acknowledged *ne plus ultra* of human foods—beans (and potatoes) pack 34 percent as much punch as a protein source; corn packs 41 percent, cow's milk 60 percent, and beef 69 percent. The peculiar nutritional talent of the bean, however, lies in its ability to complement the cereal proteins, such as those of corn. Beans are generally unimpressive sources of the sulfur-containing amino acids, such as methionine and cystine, but are excellent sources of lysine; corn, in contrast, is notoriously lysine-deficient, but contains adequate concentrations of methionine and cystine. Appropriately, our word *bean,* like the German *bohne* and the Dutch *boon,* derives from an ancient Greek verb meaning to eat.

On the other hand, "not worth beans" has meant utterly valueless since the thirteenth century, which, biochemically, doesn't seem fair. Better is "full of beans," which has meant high-spirited and energetic since the 1870s, or plain "bean," which has meant a five-dollar goldpiece since 1859. There's also "jelly bean," which has meant a dimwit since 1915. Real jellybeans, however, are made of sugar, corn syrup, and starch, and are out of the sphere of gardeners.

LETTUCE

Lettuce consumption in this century has been steadily on the rise. Americans these days each munch their way through thirty pounds of lettuce a year, over a fivefold increase since the turn of the century, and the men of a recent Armed Forces food preferences poll ranked green salads above such traditional fighting foods as meat, potatoes, and ice cream. Lettuce-eating on this grand scale, though healthfully thinning, may have its drawbacks. A dire Elizabethan anti-lettuce warning states, "The plentifull and dayly eating of the Lettuce by married persons is very incommodious and noysome to them, in that it not only doth diminish the fruitfulness of children, but the children often borne do become idle foolish and peevish persons." Such a prognostication suggests that the salad craze is behind all problems with the younger generation, which—unlike one's upstanding own—has had everything handed to it on a silver platter and nonetheless is going to the dogs. The young are said to be in their salad days, a phrase coined by Shakespeare, and a time when one is perversely lacking in *lettuce,* a slang term meaning money since the 1940s, from the even earlier American *green stuff.*

Lettuce, the non-financial kind, is a member of the daisy family (Compositae), and most likely originated in Eurasia and the Mediterranean area. The primordial lettuce plant is believed to have been *Lactuca serriola,* prickly lettuce, a wild biennial that has managed to make itself at home over most of the temperate world. Prickly lettuce is considered a pest in pastureland, since cattle who misguidedly eat it develop emphysema, but its more obnoxious attributes were not passed on to its edible descendant, *L. sativa,* the cultivated or garden lettuce. The earliest cultivated varieties had lost their spininess and apparently put out their large leaves from tall central stalks, somewhat like the seed stalks shot up by modern bolting lettuce, or those of Chinese stem lettuce or celtuce.

The Greeks, who grew *L. sativa,* referred to their lettuce as *asparagus,* a term then indiscriminately applied to any tall spike-like plant. Greek lettuce was eaten flavored with saffron and olive oil, and green pots of it were carried through the streets during the springtime Festival of Adonis. The Greeks passed lettuce on to the Romans, who disseminated it throughout their rapidly expanding empire. It became customary among the Romans to precede their gargantuan banquets with refreshing lettuce salads, in the belief that lettuce enhanced the appetite and relaxed the alimentary canal, in preparation for the onslaught of such dishes as camel's heels, flamingo's tongues, sow's teats, stuffed warblers, and sea anemones smothered in fermented fish sauce. Lettuce was also occasionally eaten cooked or pickled. The famous cookbook of Apicius (whose elaborate dinners left him overwhelmingly, and perhaps suicidally, in debt) includes a recipe for a purée of lettuce leaves and onions; Columella, the first-century agricultural writer, describes a sauerkraut-like concoction of lettuce leaves steeped in brine and vinegar.

It became customary among the Romans to precede their gargantuan banquets with refreshing lettuce salads, in the belief that lettuce enhanced the appetite and relaxed the alimentary canal.

The Roman lettuce also had its medicinal uses, primarily as a mild sedative, which clinical application has survived into modern times. The scientific name for lettuce, *Lactuca,* is derived from the Latin *lac*, "milk," from the milky sap that oozes from the cut stems. This sap is technically a latex, a water-based liquid contained in special internal channels, variously manufactured by such plants as the rubber tree; the sapodilla tree, whose dried latex, chicle, was the basis of the first chewing gums; the lettuce; and the dandelion. Latex contains numerous long-chain hydrocarbon polymers, some of which possess the desirable property of elasticity—that is, they snap back into shape after being stretched out. During World War II, Russia, deprived of classic natural rubber, made an acceptable substitute from dandelion latex; and lettuce latex, given a little chemical time and effort, could conceivably yield, if not a cost-effective bicycle tire, at least an occasional rubber band. The chemical building blocks of natural rubber are branched five-carbon molecules called terpenes, which are the basis of many plant essential oils, such as those of geranium, coriander, and mint.

In lettuce latex, particularly that of the wild *L.*

virosa, terpene-based alcohols potent enough to be nicknamed lettuce opium elicit the soporific effect. Accordingly, the Roman historian Tacitus snacked on lettuce each night before going to bed. Dried lettuce juice was used as a sleep-inducer in medieval England, occasionally mixed with henbane and poppy for additional sedative oomph, and dried lettuce juice or lettuce teas were used similarly in colonial America. A sedative drug known as lactucarium, still prepared from wild lettuce extracts, was used in hospitals up through the Second World War. Lettuce juice was also recommended, externally, for dimness of vision, scorpion and spider bites, and poison ivy, and, internally, for "wamblings of the stomach."

The medicinal reputation of lettuce may have enabled it to survive the Roman departure from northern Europe, when many other Roman-introduced plants were abandoned and lost. The early English were hardly salad fans: Queen Elizabeth ate beef, mutton, and rabbit pie for breakfast, and a perfectly acceptable Tudor company dinner consisted of meats, fish, bread, and beer, with a few "Orenges" for dessert. Favored contemporary vegetables, such as they were, included lettuce, cabbage, spinach, and radishes, though "rude herbs and roots" were generally considered fit only for the starving and the poor, or, worse, " more meet for hogs and savage beasts to feed upon, than mankind." Lettuce, ominously described in Gerard's *Herball* as "a cold and moist pot-herbe" was thought to reduce lust, a doubtful virtue, and to decrease fertility. Still, the seventeenth-century French courageously consumed lettuce hearts— candied, in a sweet dish called *gorge d'ange,* or "angel's throat"—and popularized the lettuce salad, under the domineering aegis of Louis XIV (father of seven), who liked his lettuce seasoned with tarragon, pimpernel, basil, and violets. In 1699, John Evelyn, a salad fancier and enthusiastic amateur gardener, wrote an entire treatise on the joys of salad, entitled *Acetaria, A Discourse of Sallets,* in which he referred to the lettuce as a "Noble Plant" and "the principal foundation of the universal tribe of Sallets." His list of contemporary cultivars sounds almost familiar: included are cabbage,

Favored contemporary vegetables, such as they were, included lettuce, cabbage, spinach, and radishes, though "rude herbs and roots" were generally considered fit only for the starving and the poor, or, worse, "more meet for hogs and savage beasts to feed upon, than mankind."

cos, curled, and oak-leaf lettuces, plus a breed called Passion, a peculiar moniker in the context of the times. Evelyn's lettuce, while enlivening the dinner table, was still considered dampingly beneficial for morals, temperance, and chastity.

L. sativa came to America with Columbus, who planted some in the West Indies in 1493. A quick crop, it was favored by greens-hungry early explorers. Samuel de Champlain and company planted some—perhaps the first lettuce in North America—on an island in the St. Croix River in Maine in 1619, and Captain John Smith's crew followed suit some years later, planting an island garden that "served us for sallets in June and July." The first colonists arrived equipped for salad planting. Three ounces of "lettice" seed appeared on John Winthrop, Jr.'s "bill of garden seeds" among the other horticultural essentials for emigration, carefully purchased in July 1631 from Robert Hill, London grocer, "dwelling at the three Angells in lumber streete." Young John's seed list also included the seeds of beet, "Cabedg," "Carret," "Culiflower," "onyon," parsley, radish, and "spynadg," along with the now less common "hartichockes," burnet, corn salad, chicory, orach, rocket, and purslane, all for a total cost of one pound, six shillings. The seeds, along with a vast number of other useful items—gunpowder, stockings, sheepskins, bird lime, wine vinegar, and "Conserve of redd roses"—accompanied John Jr. across the Atlantic on board the *Lyon* to join his father in the colony at Massachusetts Bay. Lettuce was planted for home use in many seventeenth- and eighteenth-century kitchen gardens. Washington and Jefferson both grew and served it, though apparently not always in accepted gourmet fashion; one ungrateful guest at Mount Vernon recorded critically that the salad lacked olive oil.

In 1806, according to Bernard M'Mahon's *Gardener's Catalog,* Americans were choosing among a mere six cultivars of lettuce; by 1828, Grant Thorburn had upped the count to thirteen, and by the 1880s, there were well over a hundred. Early catalogs and listings featured all four of the lettuce types available today; head or cabbage lettuce, butterhead, looseleaf, and cos or romaine. Heading lettuce, *L. sativa* var. *capitata,* became the universally preferred lettuce in the late nineteenth century; Burpee's 1888 *Farm Annual* listed twenty-

COMPOUND SALLET

From Gervase Markham, The English Housewife, *1631*

♦ Your compound Sallets are, first the young Buds and Knots of all manner of wholesome Herbs at their first springing; as red Sage, Mint, Lettice, Violets, Marigolds, Spinage, and many others mixed together, and then served up to the table with Vinegar, Sallet-Oyle, and Suger.

four different head lettuces, four cos cultivars, and three looseleafs; Vilmorin-Andrieux in 1886 described fifty-six head lettuces, seventeen cos types, and four looseleafs under the name "small or cutting" lettuces.

Heading lettuce is a newcomer to the vegetable world, appearing on the scene in medieval times—and perhaps only with difficulty then, since a surviving horticultural hint from the 1570s suggests trampling on young seedlings to encourage proper head formation. Advice of the same ilk outlined a means of producing "Odiferous Lettice" by embedding the seed in the seed of a citron, and more flavorful lettuce by watering nightly with sweet wine. If tried, this last technique must not have worked, since the dense-packed heads of leaves seem to have been relatively tasteless almost from their inception. Virginian John Randolph, Jr., in his early eighteenth-century *Treatise on Gardening,* says of them: "This sort of Lettuce is the worst of all kinds in my opinion. It is the most watery and flashy; does not grow to the size that many of the other sorts will do, and very soon runs to seed." In the eighteenth century it seems to have been used mostly for soups, like the solid-headed cabbages it so closely resembled. Thomas Jefferson, who grew fifteen different kinds of lettuces in the course of his exemplary gardening career, planted both a large heading variety called Dutch Brown and a miniature called Tennis-ball. Tennis-ball lettuces were particularly favored at nineteenth-century dinner parties, served whole as individual salads.

Tasteless or no, head lettuces now dominate the American market, pale cannonball-like vegetables relentlessly shipped east from the lettuce fields of California.

Tasteless or no, head lettuces now dominate the American market, pale cannonball-like vegetables relentlessly shipped east from the lettuce fields of California. (California produces some 70 percent of this country's lettuce, a crop that totals over three billion heads a year.) An earlier and less aggressive form of var. *capitata* is the butterhead lettuce, which forms soft, loose, flop-eared heads around hearts of such tender consistency that early eaters were reminded of butter. Most familiar of butterhead lettuces are the Boston lettuce and Bibb, or Kentucky limestone, lettuce, developed by Kentuckian John J. Bibb and served by the dedicated at Kentucky Derby breakfasts.

Cos, or romaine, lettuce, *L. sativa* var. *longifolia,* also forms an upright "head" around a central bud or heart, but the romaine heads are loose, long, and cylindrical. The individual leaves are elongated ovals, reminiscent of kitchen tasting spoons, and one story holds that Socrates drank his lethal dose of hemlock from such a "spoon" of romaine. The common name *romaine* is a corruption of *Roman*; the even older name *cos* is taken from the Greek island of Cos, where the Romans originally obtained these mildly tangy leaves. The lettuce arrived in France in the 1300s, when the papacy, after much ill-advised meddling in earthly politics, abruptly relocated from Italy to French Avignon, bringing their garden vegetables with them. In France, the Roman lettuce came to be known as romaine, and became so popular that outside the country it was called Paris lettuce. An old cultivar called Paris White Cos is still available today. Romaine lettuce is said to have reached China in the seventeenth century, when the ruling emperor demanded as tribute the choicest vegetable grown in each of his subject domains. He received, among other offerings, beets, scallions, spinach, and romaine lettuce, known in the Far East as the wine vegetable because it seemed to taste like wine.

Tastier yet are the looseleaf lettuces, *L. sativa* var. *crispa,* a mixed bag of nonheading greens whose flat, frilled, or double-ruffled leaves in bright green, dark red, and bronze are ornamental as well as delectable in the salad bowl. These are known as cut-and-come-again lettuces because, if consistently picked, new leaves will continue to be produced throughout the summer growing season. Prominent among them is the perenially popular Black-Seeded Simpson, a direct descendant of the Early Curled Simpson introduced by grower A. M. Simpson in 1864. Older are Oak Leaf lettuce, named for its lobed oak-leaf-shaped leaves, and an entrancing breed called Deer Tongue, whose leaves, unlike the tongues of real deer, are sharply triangular. In 1973 a modern-style looseleaf lettuce cultivar was awarded the first patent—strictly speaking, a Plant Variety Protection Certificate—ever awarded by the federal government to a plant seed product. The recipient was Burpee's

Green Ice, a deep green, crinoline-like lettuce of "outstanding taste and looks."

The advantage of the heading lettuces over their less compact relatives is their transportability. Virtually indestructible, heading lettuces—now tactfully known as crispheads—survived even turn-of-the-century transcontinental shipping, nine days from California to New York by rail. Iceberg, whose hard, cold, and pale image typifies the shipping lettuce—Ogden Nash described it as "a globe of frozen insipidity"—came on the market in 1894. It was outstripped in the 1920s by the brown-blight-resistant Imperial, which in turn gave way in 1941 to the Great Lakes cultivars, strains of which still dominate the head lettuce business today. Ironically, Americans chose to embrace the salad just as the available lettuce, in terms of taste and texture, began to go sharply downhill. On the other hand, early neglect of salads was due in large to the unavailability of any lettuce at all. Lettuce, characteristically, is perishable; impossible to pickle, preserve, dry, or stash in the root cellar, lettuce was off-limits to the average family for much of the year. For some, this green-less situation was rectified by the boom in greenhouse building in the 1830s and 40s that followed the invention of efficient hot-water central heating systems. Fruits, especially grapes, and vegetables were raised year-round in the new greenhouses, but because of high prices such produce was generally indulged in only by the upper classes. By the turn of the century, however, Boston alone had two hundred acres of vegetables—lettuce, radishes, tomatoes, and cucumbers—under glass, and from the West a combination of refrigerated railroad cars and rock-solid head lettuce cultivars supplied off-season vegetables to East Coast citizens of less fortunate climates and incomes. In 1896, Fannie Merritt Farmer, doyen of the measured cup and the level teaspoonful, observed that "salads, which constitute a course in almost every dinner, but a few years since seldom appeared on the table."

The common name *romaine* is a corruption of *Roman;* the even older name *cos* is taken from the Greek island of Cos, where the Romans originally obtained these mildly tangy leaves.

While salad lettuce appears in colors ranging from ruby-red to pearl-white, most of it is green, a color associated with the pigment chlorophyll. Chlorophyll is a complex ring compound with a central magnesium ion, chemically somewhat similar to the iron-toting heme molecule of human and animal red blood cells. The most common form of chlorophyll in higher plants is chlorophyll *a*, an attractive bright bluish-green pigment. Substitution of an oxygen molecule for a crucial pair of hydrogens converts *a* to chlorophyll *b*, olive-colored, present in about half the amount of chlorophyll *a*. The chlorophylls, along with the carotenoid pigments, absorb energy from sunlight and use it, via the photosynthetic pathway, to convert unprepossessing carbon dioxide and water into sugar. Leaves, such as those of lettuce, are specialized for this purpose, large flat surfaces crammed with chlorophyll molecules, the better to soak up the sun.

The first of Ryan's brainchildren was a chlorophyll-based toothpaste—brought out by Pepsodent in 1950 as the green Chlorodent—followed by a chlorophyll-flavored dogfood.

The green pigment received its biggest publicity boost to date in the 1950s, when creative advertising convinced the American public, always obsessed with smells, of the unparalleled efficacy of chlorophyll as a general-purpose deodorizer. The preparation involved was a water-soluble form of the pigment developed by a respectable Finnish-born doctor, Benjamin Gruskin, just before World War II. The medical potential of Gruskin's soluble chlorophyll, under the trade name Chloresium, was investigated in the 1940s by the U.S. Army. It was found to enhance wound healing, but the results were not profound enough to interest any of the major drug-manufacturing companies. They did, however, interest a double-barrelled Irishman named O'Neill Ryan, Jr., who took out a "use" patent on Gruskin's chlorophyll in 1945. The first of Ryan's brainchildren was a chlorophyll-based toothpaste—brought out by Pepsodent in 1950 as the green Chlorodent—followed by a chlorophyll-flavored dogfood. Over the next two years, spurred on by the growing ranks of chlorophyll consumers, diverse businesses entered the marketplace with chlorophyll chewing gum, mouthwash, deodorant, cigarettes, soap, shampoo, skin lotion, bubble bath, popcorn, diapers, sheets, and socks, and Schiaparelli even produced a chlorophyll cologne. Plans were in the works for chlorophyll salami and chlorophyll beer, when the Food and Drug Administration rained all over the green parade, announcing that there was "no conclusive evidence" that chlorophyll had any deodorizing effect whatsoever. The *Journal of the American Medical*

Association then caustically pointed out that goats, which practically live on chlorophyll, smell dreadful; and in 1953 a report appeared in the *British Medical Journal* detailing the work of a Glasgow University chemist who had systematically tested the effects of chlorophyll on such undesirable effluvia as skunk, onion, garlic, and human body odor, and found it a flop. Chlorophyll products subsequently petered out, and chlorophyll returned to its everyday function.

The more sunlight leaves get, the more chlorophyll and carotenoid pigments are accumulated to cope with the energy input, and the darker in color the leaves become. The outer leaves of head lettuces are thus the darkest and greenest, while the inner light-deprived leaves are pigmentless and pale. The darker leaves are the most nutritious, since they are most heavily packed with the carotenoid beta-carotene, a precursor of vitamin A. Green leaves usually contain one carotenoid molecule for every four to five molecules of chlorophyll, a ratio that supplies looseleaf lettuce with a substantial nineteen hundred International Units of vitamin A per one hundred grams (about ten leaves). Butterhead lettuce contains only half that, and the largely light-shielded crisphead lettuce only one-sixth as much. Lettuces, in similarly decreasing order, are also sources of calcium and vitamin C.

Other than that, the lettuce leaf consists mostly of water, about ninety-five percent by weight. This water is what makes lettuce crisp: cells high in water bulge turgidly against each other, producing the crunchy texture so desirable in fresh lettuce leaves. Conversely, since lettuce is so internally water-logged, it is particularly subject to water loss and wilting. To prevent this sad happening, modern lettuce-eaters often store their leaves in refrigerator crispers, confined compartments that maintain an atmosphere of high humidity.

Best, of course, is to pick it, fresh and crisp, straight out of the garden, an occupation that has improved in status since the seventeenth century, when "to pick a salad" meant to indulge in a meaningless and trivial task. Most pickers today consider themselves solidly worthwhile and rank salad-picking, in the horticultural scheme of things, somewhere comfortably between making hay while the sun shines and letting the grass grow under one's feet.

The lettuce leaf consists mostly of water, about 95 percent by weight. This water is what makes lettuce crisp: cells high in water bulge turgidly against each other, producing the crunchy texture so desirable in fresh lettuce leaves.

CELERY

Celery tonic, believe it or not, is still around. Nowadays it's called Cel-Ray and is made by Canada Dry, but it's still basically the same brew—soda water flavored with crushed celery seeds—that went on the market in Brooklyn in 1869 as Dr. Brown's Celery Tonic. Dr. Brown was at the forefront of the nineteenth-century celery craze, an enthusiasm that eventually produced not only celery soft drinks, but also celery gum, celery soup, and Elixir of Celery, touted as a treatment for nervous ailments and popular enough to be offered in the 1897 Sears, Roebuck catalog.

Celery was also in vogue on the menu. For much of the last century, it was considered a high-status vegetable, since it was too steeply priced for the economical tables of average citizens. Cost was related to production: celery cultivation was particularly laborious because the plants were routinely blanched, a whitening and sweetening process that involved piling dirt around the developing stalks to block exposure to sunlight. The palely elegant results were served, in accordance with their aristocratic image, in special celery stands, tall, footed vases of cut-glass or silver that loomed expensively over the dinner table. The introduction of self-blanching celery varieties—Burpee's Golden Self Blanching Celery came on the market in 1884—brought the crunchy stalks within reach of the ordinary pocketbook. It was soon immensely popular across the social board; a late-nineteenth-century French tourist peevishly wrote that the celery-obsessed Americans "almost incessantly nibble from the beginning to the end of their repasts." Inevitably, as celery came within the common clutches, it lost its prestige, and by the 1890s had been demoted from the towering vase to an unobstrusive flat dish—the celery "yacht" or "boat."

The subject of all this social heartburn is a member of the carrot family, Umbelliferae, a cousin of parsnips, parsley, caraway, coriander, dill, and fennel. Its scientific name, *Apium graveolens,* derives from the Latin *apis,* or "bee," because bees go dotty over its tiny white flowers. Early documents refer to celery (and, confusingly, parsley) as ache (pronounced ash), from which developed the Old English small-ache and smallage. The related lovage—a very large and strong-flavored celery-like plant—comes from the same linguistic root, originally love-ache or love-parsley.

Celery is believed to have originated in Eurasia, where, in wild form, it established itself in the marshlands adjacent to the seacoast. The ancients collected it for medicinal use, and one linguistic theory holds that our common name *celery* derives from its remedial reputation—from the Latin *celer,* meaning quick-acting or swift. The Egyptians used celery stalks to treat impotence; the Romans used them to treat constipation, and wore them to alleviate hangover: Apicius, who doubtless suffered many in the course of his expensively decadent career, recommended that morning-after victims "wear a wreath of celery round the brow to ease the pain." The Greeks associated celery with funerals, used it to bedeck tombs, and coined the ominous saying "he now has need of nothing but celery" to mean imminent and unavoidable demise. In the Middle Ages, celery was used as a laxative or diuretic, as a treatment for gallstones, and as a palliative for wild animal bites. Madame Pompadour, with celery's rumored aphrodisiac effect in mind, fed Louis XV on celery soup. Therapy in the eighteenth and nineteenth centuries included celery (in tea) for digestive upsets and insomnia, and (in conserve) for chest pains. Early ships' crews—Captain Cook's among them— collected and consumed it (raw) as an antiscorbutic.

The Egyptians used celery stalks to treat impotence; the Romans used them to treat constipation, and wore them to alleviate hangover: Apicius, who doubtless suffered many in the course of his expensively decadent career, recommended that morning-after victims "wear a wreath of celery round the brow to ease the pain."

Modern medical interest in celery centers around its content of chemical compounds known as psoralens or furocoumarins. These compounds, present in appreciable amounts in parsnips, celery, and parsley, are potent photosensitizers—that is, they increase the sensitivity of the skin to sunlight. In combination with ultraviolet irradiation, psoralens have been used to treat vitiligo—a skin depigmentation condition—and psoriasis, a chronic and miserable inflammatory skin disease. In retrospect, however, it seems that the psoralens may be yet another case of a treatment worse than the disease. Recent indications are that they are photocarcinogens: psoralen-treated patients appear to have an increased incidence of skin cancers. The psoralens are believed to bring this about by insinuating themselves into the strands of the skin cell DNA, where, in the presence of light, they form damaging crosslinks with the normal pyrimidine bases. In healthy celery, psoralens are present only at low levels and pose no threat to the human hide. Sick celery, however, is a different matter: the celery plant produces psoralens in response to microbial invasion, and diseased plants thus possess ten to one hundred times more of these molecules than their germ-free relatives, potentially dangerous concentrations. Parsnips are worse, even in the pink of health relative hotbeds of psoralens. Consumption of 0.1 kg of parsnip (a mere 3½ ounces) necessarily involves engulfing 4–5 mg of assorted psoralens, a risky quantity unless one lives in a lightless cellar.

The chief psoralens in celery, bergapten and xanthotoxin, are capable even at normal low levels of inducing severe allergic reactions, frequently afflicting celery harvesters and growers, and ranging in severity from hives to outright anaphylactic shock. Similarly implicated in celery allergies is apiol, the essential aromatic oil of the celery plant, which is most highly concentrated in the spicy seeds. The major constituent of apiol is a terpene compound called limonene, also present in citrus fruits and various mints. (Limonene is the *bête noire* of the fruit juice industry: it turns distressingly bitter in processing, and has therefore inspired a scientific and industrial quest for low-limonene oranges and grapefruits.) The allergic reaction is said to be exacerbated by exercise, a good argument for lying quietly on the couch while chewing celery stalks.

In the culinary sense, apiol has been in demand for centuries. As well as in the seeds, the oil is concentrated in cavities between the cells of the leaves, which is why leafy celery tops are often used to flavor soups. Celery seed was used as a condiment in ancient Rome—Apicius, for example, sprinkled it in a sauce to be served with wild boar. The Romans preferred the tougher and stronger-flavored wild celery to the milder cultivated variety, but seem, at least sporadically, to have eaten both. With the decline and fall of Rome, celery in general fell from public view. It survived the Dark Ages as a medicinal, and had some appeal as an art form—a stylized celery leaf in solid stone appears among the gargoyles on Gothic cathedrals—but was resurrected as an edible, according to food historians, only in the sixteenth century. Western Europe acquired its table "Sallery" by way of Italy, where hefty wide-stalked varieties similar to those around today were originally developed. The Reverend William Turner chattily mentions Italian celery in his *Herbal* of 1538: "The first I ever saw was in the Venetian Ambassador's garden in the spittle yard, near Bishop's Gate Streete." (Spittle, less repulsive than it sounds, was an early English term for garden spade.) Sixty years later, however, the authoritative Gerard ignores celery altogether, and in 1629, John Parkinson describes it as a rarity. By Parkinson's time, the culinarily adept French were in the midst of developing an elaborate celery cuisine. A recipe of 1659 describes a dish of celery cooked with lemon, pomegranates, and beets; celery hearts—the fused base where the stalks meet—were esteemed; and the leafy stalks themselves were eaten as delicacies with dressings of oil and pepper. The practice of blanching celery began in France in the reign of Louis XIV, under the direction of Jean de la Quintinie in the Sun King's legendary kitchen garden at Versailles. Blanched celery, sweeter, tenderer, and considerably whiter than the sun-soaked norm, quickly became the dietary standard. The earliest blanching techniques, also applied to rhubarb and endive, involved "earthing up" or piling soil around the growing stalks; later growers employed earthenware blanchers, large bell-jar-shaped pots with removable lids.

Celeriac, historians believe, appeared out of the blue sometime in the seventeenth century, as a chance mutation in an unknown garden. Known as knob, or turnip-rooted, celery, celeriac features a starch-swollen lower stem (not root). Its scientific name, *A. graveolens* var. *rapaceum,* reflects the necessary harvesting technique—unlike the more docile garden celery, *A. graveolens* var. *dulce,* celeriac clings to Mother Earth and has to be ripped from the ground by force. Historically, it seems to have caught on best in Germany and France, where it was usually boiled. Stephen Switzer, an eighteenth-century English gardener and garden writer, obtained celeriac from an importer of "curious seeds," who in turn had procured it from Alexandria. Switzer grew some, and included it in his 1729 treatise titled "A compendious method for the raising of Italian brocoli, Spanish cardoon, celeriac, finochi and other foreign vegetables," which suggests that it was at that time a vegetable oddity. John Randolph of Virginia, however, included both celery and celeriac on his recommended list of plants for American kitchen gardens.

The colonists brought celery with them to America, repeatedly and unsuccessfully, since the climate, north and south, never seemed to suit it. The gardeners of Massachusetts Bay reported that their "Celary" roots rotted over the winter, and Thomas Jefferson recorded similarly unhappy results at Monticello. Somebody managed to grow it, however: Bernard M'Mahon lists four varieties common to American gardens, in 1806, though it certainly never reached such heights of commonness as the bean, the pea, and the onion. Commercial celery didn't appear until mid-century, the successful stock imported from Scotland in 1856. Other sources claim the crucial seed arrived with a green-thumbed Dutch immigrant in the 1870s. All agree that celery as a commercial money-maker was first grown in Michigan, which, with California and Florida, remains a leading producer of celery nowadays. (Top dog is California, which turns out 60 percent of the national crop.) Burpee's 1888 catalog offered ten varieties of celery, including the Incomparable Crimson and the White Walnut, so named for its rich, nutty flavor, plus celeriac or turnip-rooted celery, which, Mr. Burpee adds truthfully in parentheses, is really shaped like an apple. Until the 1930s most celery in this country was sold blanched and white, but the most likely denizen of the supermarket vegetable bins today is a variety called Pascal, which is green.

Celery is still characteristically eaten crisply raw, much as advised by the omniscient *Ladies' Home Journal* in 1891, which directed:

With the decline and fall of Rome, celery in general fell from public view. It survived the Dark Ages as a medicinal, and had some appeal as an art form—a stylized celery leaf in solid stone appears among the gargoyles on Gothic cathedrals—but was resurrected as an edible, according to food historians, only in the sixteenth century.

"Celery should be scraped and washed and then put in iced water, to be made crisp, at least an hour before it goes on the table. It is now served in long, flat glass dishes. It should be put on the table with the meat and other vegetables, and is to be removed before the dessert is served."

Celery is 95 percent water, which contains no calories, but it still manages to offer a good dose of minerals and of vitamin A.

In this form, celery contains about seven calories per average stalk—not, as wishfully preached by dieters of the last decade, "negative calories," meaning that it takes more calories in energy to eat than are present in the food in the first place. Celery is 95 percent water, which contains no calories, but it still manages to offer a good dose of minerals and of vitamin A. Celery is also a principal ingredient of the famous Waldorf salad—along with apple, walnuts, and a dollop of mayonnaise—and in *Greene on Greens* master vegetable chef Bert Greene offers ten pages of cooked celery cuisine, including a mouthwatering celery in brandied cream sauce.

Celery stalks are leaf petioles rather than true stems—structural equivalents of the "stems" that attach conventional leaves to a main branch. They tend to be finicky to grow, distressingly sensitive to temperature and soil composition. Modern celeries, descendants of an ancestral seaside plant, still prefer a boggy, acidic soil of the kind known inelegantly in the celery trade as "muck." Most commercial celery spends the first two to three months of life in a temperature-controlled greenhouse, before being painstakingly transplanted to the great outdoors. Reaching the edible stage then takes another one hundred days or so, depending on soil, weather, and celery cultivar. In an attempt to simplify matters, researchers at Plant Genetics, Inc., in Davis, California, have been developing "synthetic seeds" for celery and similar high-cash crops. The synthetic seeds, actually pre-germinated embryos encased in a biodegradable capsule, are designed to give the infant celery a developmental headstart in the field. The embryos are derived from individual celery stem cells, grown up in tissue culture, and encapsulated in a protective jelly-like coat of sodium alginate, a polymer extracted from seaweed. The result looks, says one science writer, "like cod-liver oil pills." The pills contain populations of genetically identical future plants, picked from the cream of the parental celery crop, and destined, as far as scientists can yet control such things, for instant agricultural success.

Production of hybrid celery seed by conventional breeding techniques is a notoriously tedious and difficult process, for which reason growers are interested both in large-scale generation of genetically superior clones, as in synthetic seeds, and in the selection of new and better celery variants, most recently by means of the process of somaclonal variation. So far, the technique has been used to produce a stringless celery and a sweeter carrot—both aimed at the vegetable snack-foods market—and a tomato with an exceptionally high solids content, the better to make condensed soup and ketchup. In the works: a naturally butter-flavored popcorn, a more powerful garlic, a mintier mint, and a right-off-the-tree caffeine-free coffee. Nonsynthetic celery seeds, of which there are five hundred to the ounce, are still used as a seasoning, much of it imported from India, where certain celery cultivars are raised specifically for production of seed. It's a mild and appealing taste, but alternativc uses are more exciting. Medieval magicians, the story goes, tucked celery seeds in their shoes in order to fly.

CARROTS

The first carrots, botanists believe, came from Afghanistan and were colored purple. These original roots, scrawny, highly branched, and unpromising, belonged like their plump cultivated descendants to the species *Daucus carota*, a member of the Umbelliferae family. The ancestral carrot probably looked very much like Queen Anne's lace, the ubiquitous "wild carrot" of present-day fields and roadsides. The cultivated Greek and Roman carrots were likely still branched—a characteristic referred to by modern breeders as "a high degree of ramification"—but were certainly larger and fleshier than their wild relatives. Experiments in the late nineteenth century by French seedsman Henri Vilmorin demonstrated the relative ease with which primitive farmers could have developed the cultivated carrot. Starting with a spindly rooted wild species, Vilmorin was able to obtain thick-rooted equivalents of the garden carrot in a mere three years. The conical root shape characteristic of carrots today seems to have shown up in the tenth or eleventh century in Asia Minor, and reached Europe in the twelfth century by way of Moorish Spain.

The primitive purple, violet, red, and black carrots owed their color to anthocyanin, a pigment that dominated the carrot world until approximately the sixteenth century, when a pale yellow anthocyanin-less mutation appeared in Western Europe. It thus must have been an anthocyanin carrot that Achilles' soldiers munched (quietly) inside the Trojan Horse "to bind their bowels," and that Greek doctors on the home front used to concoct an aphrodisiacal potion called Philtron. In the same tradition, devious Roman soldiers boiled carrots in broth to release the sexual inhibitions of their female captives, and the Emperor Caligula, who had a fun-loving streak, once fed the entire Roman Senate a feast of carrots in hopes of watching them run sexually amok.

The purple carrot quickly fell from favor with the advent of the yellow and, later, orange varieties. The purples, though tasty, turned an unappetizing brownish color in cooking, which put off even the most stolid of sixteenth-century chefs. The aesthetically appealing orange carrot was a seventeenth-century development of the Dutch, who were the dominant European carrot breeders. From this orange original, the Dutch growers soon produced the deeper-colored Long Orange, a hefty carrot intended for winter storage, and the smaller and sweeter Horn. The Horn was further fine-tuned to yield, by the mid-eighteenth century, three breeds of orange carrot varying in earliness and size: the Late Half Long, the Early Half Long, and the smaller Early Scarlet Horn. Collectively, these two-hundred-year-old Dutch carrots are the direct ancestors of all orange carrots grown today.

The orange color indicated the presence of carotene, a precursor of vitamin A. The carrot is a prime source of vitamin A, containing 7,930 International Units per average-sized root. A single carrot, therefore, more than fulfills the adult daily vitamin A requirement. Even better than the average carrot is the "supercarrot," the outcome of thirty years of crossbreeding by University of Wisconsin plant geneticist Warren Gabelman. Gabelman's carrot, genetically constructed for enhanced carotene synthesis, contains 40 percent more beta-carotene than the run-of-the-mill root, and has been given the descriptive name Lucky's Gold. As well as giving carrots their orange and superorange, it's carotene that gives cream its rich yellowish hue. Cows fed on the champion Dutch carrots were said to yield the richest milk and the yellowest butter in seventeenth-century Europe, which in turn was held to be responsible for the famous rosy-cheeked Dutch complexion. Colonial buttermakers, starting with less well-fed cows, often colored their butter after the fact by adding carrot juice to the churn. (Nowadays the yellow in butter comes from annatto, a natural dye extracted from seeds of *Bixa orellana*, a tropical tree.)

Cows fed on the champion Dutch carrots were said to yield the richest milk and the yellowest butter in seventeenth-century Europe, which in turn was held to be responsible for the famous rosy-cheeked Dutch complexion.

If you eat enough carrots, you can overdose on carotene and turn yellow, a condition rare, but possible, in human beings. The syndrome is known clinically as carotenemia and results from the deposition of carotene pigments in the subcutaneous fat, which is located just beneath the skin. Though visually spectacular, carotenemia seems to be physically harmless, and disappears within a few weeks if the victim lays off carrots. A variation on the yellow theme is an orange condition called carotenemia-lycopenemia, which occurs when overindulgence in carrots is accompanied by overindulgence in tomatoes. The color results from the simultaneous accumulation of carotene (yellow) and lycopene (red), which mix, just like fingerpaints, to make orange.

Carrots are also high in minerals—potassium, calcium, and phosphorus—and low in calories. Americans each eat about ten pounds of them a year, to the tune of $100 million total. Carrots thus make it into the vegetable Top Ten, seventh in consumer popularity behind lettuce, tomatoes, onions, celery, cabbage, and sweet corn, but well ahead of spinach, turnips, and beets. For all their caloric skimpiness, carrots are high in sugar, possessing some seven grams of carbohydrate per medium-sized root. For this reason, carrots traditionally were eaten in desserts. A recipe for a carrot conserve survives in the *Menagier de Paris*, a collection of housekeeping hints compiled in the fourteenth century by a concerned Frenchman for the edification of his fifteen-year-old bride. The conserve, in which carrots are referred to as "red roots," also contains green walnuts, mustard, horseradish, spices, and honey.

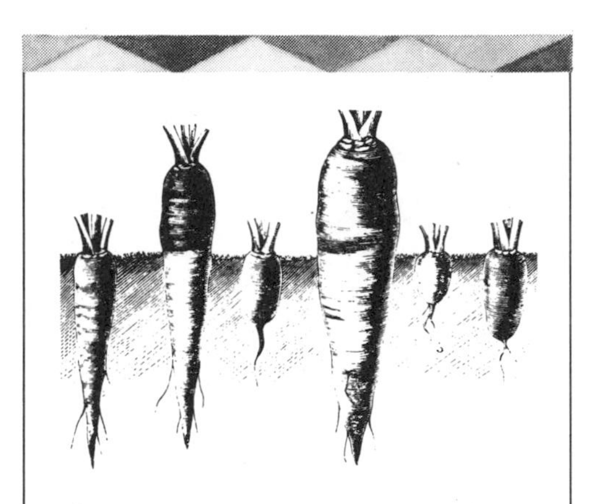

Carrots similarly sweeten over winter: the best, says Vermont grower John Page, spend the winter under two feet of leaves and hay, and are harvested in the spring.

Later carrot cuisine featured pies, cakes, tarts, and carrot puddings, in which the cooked vegetable, mashed, was mixed with eggs, cream, sugar, nutmeg, cinnamon, and sherry wine. The Irish, who doubtless found carrots a welcome relief from a relentless diet of potatoes, called them "underground honey." The English and Americans used them to make carrot jam and, occasionally, carrot wine. The natural sugar concentration made the carrot an excellent subject for fermentation; one early calculator estimated that one acre of carrots should yield 168 gallons of spirit. During the lean years of World War II, carrot popularity boomed. Under the urging of "Dr. Carrot," invented by the

British government to encourage vegetable-eating on the home front, innovative housewives produced carrot toffee, carrot marmalade, and a nonalcoholic drink called Carrolade made from crushed carrots and rutabagas.

ARSNIP WINE

From Mackenzie's 5000 Recipes, *1829*

♦ To 12 pounds of parsnips, cut in slices, add 4 gallons of water; boil them till they become quite soft. Squeeze the liquor well out of them, run it through a sieve, and add to every gallon 3 pounds of loaf sugar. Boil the whole three quarters of an hour, and when it is nearly cold, add a little yeast. Let it stand for ten days in a tub, stirring it every day from the bottom, then put it into a cask for twelve months: as it works over, fill it up every day.

Even sweeter than carrots are the related parsnips, *Pastinaca sativa*, of which Ogden Nash, not a parsnip lover, penned:

The parsnip, children, I repeat,
Is simply an anemic beet.
Some people call the parsnip edible;
Myself, I find this claim incredible.

The parsnip has been on its way out since the nineteenth century, when, like the Jerusalem artichoke, it was ousted by the more versatile Irish potato. Pre-potato, however, it was a garden staple: the rich ate parsnips in cream sauce, the poor ate them in pottage; John Gerard and Sir Francis Bacon both put in a good word for them, and Sir Walter Scott wrote sardonically that fine words don't butter any. Most parsnips are long, white, funnel-shaped roots. There are also round turnip-like varieties that, though introduced to the United States in 1834, never really caught on. The large meaty parsnips known as the hollow crown varieties were developed at some point in the Middle Ages; still grown today, these have a saucer-shaped depression at the top of the root (the "hollow crown") from which sprout the stems and leaves.

The sugary parnsip has been boiled down into syrup and marmalade, and, with the help of a little yeast, brewed into beer and parsnip wine. One of Mackenzie's *5,000 Recipes* (1829) directs hopeful winemakers to boil twelve pounds of sliced parsnips, strain through a sieve, and add to the liquid loaf sugar and yeast. Mackenzie recommends an aging period of twelve months; modern winemakers, according to biologist (and parsnip fan) Roger Swain, opt for up to ten years, and many wine connoisseurs suggest never taking the stuff out of the cask. The sweeter the parsnip, the more efficient the fermentation process—which means that winemakers should harvest their parsnips in the spring.

Many plants, including parsnips, cabbages, and

potatoes, sweeten after exposure to temperatures under 50 degrees F (10 degrees C). As early as the first century A.D., Pliny commented on the phenomenon: "turnips are believed to grow sweeter and bigger in cold weather" and "with any kind of cabbage hoar-frosts contribute a great deal to their sweetness." Low-temperature sweetening occurs in leaves, shoots, and roots of responsive plants. The sugar content of cabbages increases 100 percent after thirty days in the cold; over-wintered parsnips contain nearly three times more sucrose by weight than their autumn-harvested buddies. Carrots similarly sweeten over winter: the best, says Vermont grower John Page, spend the winter under two feet of leaves and hay, and are harvested in the spring.

No satisfactory explanation yet exists for this sweetening process. One hypothesis suggests that the increased sugar acts as a cryoprotectant, or species of natural antifreeze. A similar explanation has been advanced to account for the increased spring sugar content of certain tree saps—notably that of the sugar maple. Unfortunately, while low-temperature sweetening does seem correlated to cold hardiness in some cases, such as the cabbage, the relationship does not hold true in others. In the potato, for example, though sugar content increases, the tubers are not rendered cold-resistant. Instead, they simply become unpalatable and unprocessable. Sugars in sweetened potatoes react with amino acids to cause intense and unsightly browning in the production of potato chips.

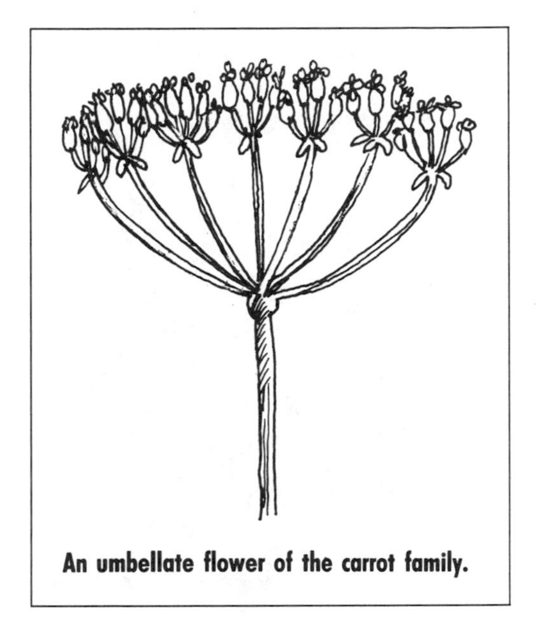
An umbellate flower of the carrot family.

Both carrots and parsnips are biennials. The starches and sugars of specialized storage root are intended to support the development of flowers and seeds in the second year of growth. Carrot flowers are lacy compound umbels on two- to six-foot stalks similar to those of Queen Anne's lace. The largest flower, which ripens first, is called the king umbel, followed by a lesser array of side umbels. Carrot foliage is similarly attractive. John Parkinson wrote of it in 1629: "The carrot hath many winged leaves . . . of a deep green colour, some whereof in autumn will turn to be of a fine red or purple (the beauty whereof allureth many Gentlewomen often times to gather the leaves, and stick them in their hats or heads, or pin them on their arms instead of feathers)"—which must have been a fine, if shortlived, sight.

The edible taproot, for which gardeners forfeit the aesthetic delights of the carrot flowers, consists of a central core of vascular tissue and an outer layer called the cortex, composed of storage tissue. In the carrot, as increasing amounts of starch are accumulated, the central core pushes outward, maintaining a two-part pattern. In contrast, in the beet, vascular and storage tissue are laid down in alternate rings, like those of a tree trunk; while in the radish and turnip, vascular and storage cells are indistinguishably intermixed. The taproots of most carrots grown in today's garden are five to eight inches long. Early field carrots were much bigger, and nineteenth-century growers boasted of two-foot roots, a foot or more in circumference at the thickest, weighing up to four pounds. Amelia Simmons in *American Cookery*, after a discussion of the superiority of the yellow carrot over the orange and the red, advises use of the "midling sized" carrot in cooking, by which she means a vegetable a foot long and two inches across at the top. She suggests it be served with hash.

Early field carrots were much bigger, and nineteenth-century growers boasted of two-foot roots, a foot or more in circumference at the thickest, weighing up to four pounds.

The carrot arrived in North America with the first settlers. The Jamestonians planted them, between tobacco crops; John Winthrop, Jr., included "Carrets" on his seed list. Carrots figured in the earliest of American seed advertisements: in 1738, one John Little offered orange carrots for sale along with his other "new garden seeds"; in 1748, "Richard Francis, Gardner, living at the sign of the black and white Harre at the South end of Boston" offered the gardening public a choice of orange or yellow carrot seeds. Jefferson grew carrots, in several colors, at Monticello. From all of these widespread colonial gardens, the carrot promptly escaped and reverted to the wild. The American Queen Anne's lace descends from such ex-cultivated escapees. The name is said to refer to Anne of Denmark, wife of James I and an expert needlewoman. Queen Anne, the story goes, in an attempt to alleviate the boredom of court living, challenged her ladies-in-waiting to make a piece of lace as fine as the flower of the wild carrot. The Queen herself, not surprisingly, won hands down, and the flower was re-christened in her name. Less romantically, it is known as bird's nest or devil's plague. In some areas, Queen Anne's lace is cultivated in its own right, for its slender white roots. In its edible version, it is known as white carrot.

The famous Danvers carrot, a dark orange, medium-long variety with exceptionally high yields, was developed in Danvers, Massachusetts, in the 1870s. It may have been a Danvers carrot, tricked out in top hat, monocle, and walking stick, that was featured among the Buckeye Phosphate advertising cards, trade gimmicks of the nineteenth century designed to publicize various useful products. The Buckeye Phosphate manufacturers touted their fertilizer through portraits of immense and elaborately costumed vegetables, accompanied by shamelessly dreadful couplets. The carrot bore the legend: "BUCKEYE Phosphate is the 'ticket for soup' Carrots raised with it all other kinds scoop."

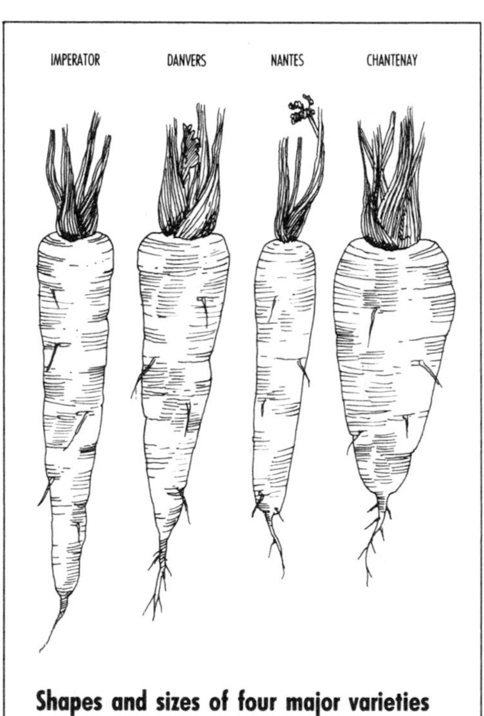

Shapes and sizes of four major varieties of carrots.

On both sides of the Atlantic, the 1870s were a banner decade for carrots, also ushering in the Nantes carrot, named for its town of origin in France. Burpee's 1888 *Farm Annual* neglected the Nantes in favor of the equally French Chantenay ("of more than usual merit") and Guerande ("of very fine quality") carrots—but reserved special praise for the homegrown Danvers, variously described as "handsome" and "first-class." Also offered were the Early Scarlet Horn and the related Golden Ball, a stumpy little vegetable that looked less like a carrot than a deep-yellow radish. Vilmorin-Andrieux diplomatically mentions both the Danvers and the Nantes varieties along with some twenty-three other carrots, in a range of shapes and sizes. Impressive among them is the English Altringham carrot, with bronze or violet roots over twenty inches long.

Garden carrots today generally belong to one of four major types: Imperator, Danvers, Nantes, or Chantenay. In general, Imperator and Danvers carrots are long and pointed; Nantes and Chantenay carrots, shorter and blunt; but the vast number of carrot cultivars available these days tend to blur distinctions. It's safe to say that there's a carrot out there for everybody.

"The day is coming," said Paul Cezanne, "when a single carrot, freshly observed in a painting, will set off a revolution." He couldn't have picked a better vegetable.

RADISHES

Today's garden radish in many circles is barely considered a food at all, which from a nutritional standpoint is only just, since the radish, barring a bit of vitamin B and a tad of iron, contains next to nothing in the way of nourishment. Instead, the peppery little balls are often relegated to the level of garnishes, most popularly in the form of red-and-white scalloped radish roses. The early radish was a much heftier proposition altogether. It was, chances are, *Raphanus sativus* var. *niger*, or what our ancestors knew as the winter-keeper, or cooking radish. The earliest cultivated varieties seem to have been black-skinned—hence *niger*—and the Greeks and Romans, who described them, mention run-of-the-mill weights in the range of forty to one hundred pounds. Ancient Jewish legend tells of a radish so immense that a fox hollowed it out and used it for a den, possibly the most creative use of var. *niger*, though there have been many. It appeared in the tomb paintings of the ancient Egyptians; was featured in Greek temple offerings, carried on solid-gold plates; and was used by the rowdy Romans as a projectile during politically unpopular public speeches. When not angrily airborne, it was eaten, either cooked like a turnip, or served raw in salad, seasoned with honey, vinegar, and salt.

The radish is believed to have originated somewhere in western Asia, just east of the Mediterranean. Though several wild species exist today, none quite fills the bill as the definitive ancestor of the present-day cultivated vegetable. Closest perhaps is *Raphanus raphanistrum*, a weedy radish collected and used in early Europe as a pot-herb. Its spicy seeds, redolent of mustard oil, are still occasionally used in lieu of mustard seeds in India. *R. raphanistrum*, like many ancient herb plants, came in for its share of unscientific lore. It was said, among other sterling virtues, to possess the ability to detect the whereabouts of witches.

The radish, like the assorted cabbages, the turnip, and the horseradish, is a member of the mustard family, Cruciferae. Its relationship to the Brassicas is close enough for successful crossing, though such cross-species matings—like those of zebras with Shetland ponies—are not the natural rule. Scientists have exploited this genetic compatibility, however, in cabbage × radish protoplast fusion experiments. In this procedure, bypassing the natural rule altogether, laboratory researchers chemically remove the tough cell wall that surrounds each plant cell, leaving naked jelly-like blobs of protoplasm called protoplasts. Protoplasts from different plants can be induced to fuse by various chemical or physical techniques—a jolt of electricity has been found particularly effective—and the resultant genetically scrambled hybrid regenerated in tissue culture to form a whole plant. The success of protoplast fusion depends largely upon the degree of relationship between the parent plants: too distant a relationship ends in genetic stalemate. Successful to date have been tomato × potato, carrot × parsley, and cabbage × radish crosses, this last more of a scientific than a horticultural success, producing inedible cabbage roots below the ground and inedible radish leaves above.

Ancient Jewish legend tells of a radish so immense that a fox hollowed it out and used it for a den.

The common name *radish* comes from the Latin *radix*, meaning root, though the edible portion of the radish, as in the turnip and rutabaga, consists of both root and hypocotyl, the starch-stuffed base of the stem. Cultivated radishes fall roughly into four major groups: besides those grown for roots, there are varieties raised for seed pods, for edible leaves, and for oil-bearing seeds. Food historian Waverly Root suggests that the oil-seed radish, *R. sativus* var. *oleifera*, may have been the first to be domesticated, grown by the ancient Egyptians before they obtained the more satisfactory oil-generating olive. Radish-seed oil persisted as a valuable commodity into Roman times; Pliny commented snidely on farmers who abandoned planting low-priced grain in favor of radishes, which yielded high-priced oil.

The radish is a biennial. The seeds are ordinarily produced the second year, their development fueled by starch stored in the bulbous root during the first. The tiny seeds are borne in bean-like pods, edible in their own right. In the seventeenth century, such radish pods, pickled, were a standard accompaniment to the dinner roast. Today, the rat-tailed or snake radish, *R. sativus* var. *caudatus*, is grown specifically for its edible pods. The roots of var. *caudatus* never amount to much, but the seed pods—called siliques—are spectacular, pencil-thin, curly, and over a foot long. The pods are eaten raw in salads, where they taste much like radish roots, or are pickled or stir-fried. Leaf radishes, *R. sativus* var. *mougri*, are grown in southeast Asia, but have attracted little interest in the root-dominated West.

Of the root radishes, *R. sativus* var. *radicula*, the small, round, cherry-sized types so popular nowadays, are relatively recent developments, first appearing in the eighteenth century. Until these came along, the cooking radish, of which there were numerous cultivars, held the limelight, though Pliny called them "a vulgar article of diet," explaining that "all radishes breed wind wonderful much, and provoke a man that eateth of them to belch." Despite these social handicaps, the radish in its heyday occupied both a substantial place on the dinner table and a sizeable niche in the medicine cabinet. It was used as a general-purpose antidote to poison, to treat viper or adder bites, to alleviate the pains of childbirth, and to remove freckles. John Gerard recommended radish roots for baldness, mashed with honey and mixed with a little powder of dried sheep's heart. William Salmon prescribed "juyce of Radishes" for deafness, at the top of a list of less appealing remedies, including the "fat of a mole, eele, or serpent," essence of bullock's gall, and distilled boy's urine.

The cooking, or winter, radish reached the Far East by about 500 B.C., where it established itself in an assortment of enormous forms that flabbergasted later Western travellers. A Dutch East India Company ship is said to have brought the first Chinese radish seeds around the Cape of Good Hope to the keeper of the physic garden of the London Apothecaries' Company sometime in the sixteenth century. Linnaeus acquired some Chinese seeds from a Swedish sea captain and grew gargantuan radishes in his Uppsala garden. Com-

John Gerard recommended radish roots for baldness, mashed with honey and mixed with a little powder of dried sheep's heart.

modore Matthew C. Perry, belatedly opening Japan in 1853–54, mentioned awesome radishes, up to a yard long and a foot in circumference. Such radishes remain a staple of Far Eastern cuisine today, and seeds of Oriental cultivars can be obtained from American seed companies. Perry's radish was most likely *R. sativus* var. *longipinnatus*, the Japanese daikon. The name translates as "long root," which it is; the average daikon runs about eighteen inches long. The Japanese, who have at least a hundred different ways of cooking their mega-radishes, also eat them raw or pickled. Pickled daikon, called *takuan*, is named for the culinarily creative seventeenth-century priest who invented the traditional method of preserving the national radish. Takuan first dried his daikons in the sun, then pickled them in rice bran and salt, pressing the mixture down with a heavy stone. Takuan's tombstone is said to be shaped like a pickling stone.

"We wish that Americans appreciated good radishes and used them as largely as do the French. For breakfast, dinner, and supper, three times a day, they are a most appetizing and wholesome relish."

The radish reached North America with the European settlers and was grown in considerable variety in colonial gardens. Peter Kalm, tooling through New York in 1749, commented on the customary Dutch breakfast, which consisted of tea with brown sugar, bread and butter, and radishes. (Dinner was meat with turnips or cabbages; supper, corn porridge and buttermilk.) At least ten varieties were grown in gardens by the late eighteenth century, of which the insatiable Jefferson grew eight: the black, the common, the English scarlet, the salmon, the scarlet, the summer, the violet, and the white. According to Amelia Simmons, the "Salmon-coloured" was the best, next-best the purple. "They grow thriftiest," she tells us, "sown among onions." (Amelia gives much shorter shrift to the pungent radish relative, horseradish: "Horse Radish, once in the garden, can scarcely ever be totally eradicated; plowing or digging them up with that view, seems at times rather to increase and spread them.") The colonists grew both the large, mild winter radishes, suitable for months of storage in the family root cellar, and the smaller, zingier, and more quickly maturing summer radishes. The earliest of these summer varieties, white and carrot-shaped, were introduced in the late sixteenth century; white and red globular forms were developed by the eighteenth, and thereafter shapes, sizes, and colors proliferated.

By the nineteenth century, Vilmorin-Andrieux listed twenty-five varieties of summer radishes, subdivided by shape into round, or "turnip-rooted," radishes (ten kinds in scarlet, white, dark violet, or dull yellow); "olive-shaped" radishes (seven kinds, including the still-popular French Breakfast, which looks less like an olive than a cork); and "long" radishes (eight kinds, including the bizarre Mans Corkscrew, a pure white radish with a foot-long, sharply twisted, zigzag root, almost impossible to pull out of the ground without breaking in two). In 1888, the Burpee seed catalog listed twenty-eight varieties of early, or summer, radishes, preceded by the reproachful sales pitch: "We wish that Americans appreciated good radishes and used them as largely as do the French. For breakfast, dinner, and supper, three times a day, they are a most appetizing and wholesome relish." Prominent among them was Burpee's yellow-skinned Golden Globe, first introduced in 1880, of which an ecstatic Missouri customer wrote that some of his were sixteen inches in circumference, or about the size of a grapefruit. Also offered were the rat-tailed radish and five varieties of winter radishes, including the famous Black Spanish (sausage-shaped) and the China Rose.

The leading radish today is probably Cherry Belle, an early, mild-flavored, round radish, with a crisp white flesh and cherry-red skin. It's a bit bigger than the average cherry, the better for carving into radish roses— but if you prefer gobbling yours whole, with salt, there are alternative garnishes. Jerry Crowley, author of *The Fine Art of Garnishing*, describes techniques for making edible roses out of beets, cucumbers, potatoes, and tomatoes—plus, for the more ambitious, onion chrysanthemums, orange elephants, turnip lilies, and watermelon whales.

CABBAGES

The cabbage, *Brassica oleracea*, is a member of Cruciferae, the mustard family, along with such other edible goodies as true mustard, radishes, and watercress. Its original wild ancestor was a seaside dweller, a native of the Mediterranean and the northern European coast. Some cabbage students believe it was sea kale, a tough, bitter, loose-leafed plant still found growing wild along the temperate seaboard; others believe the modern cultivated cabbage received genetic input from a number of assorted—and now impossible to sort out—wild relatives. Whichever the case, as of the present day some prolific great-grandaddy *B. oleracea* has spawned kale, kohlrabi, head cabbage, broccoli, cauliflower, and brussels sprouts—plus the phenomenal ten- to fifteen-foot-tall tree cabbage, stalks of which are used as rafters to support the thatched roofs of cottages in the Channel Islands. (The leaves are fed to cows.)

The cultivated cabbage has been around for thousands of years: the Greeks ate it and the Romans positively doted on it. The Romans claimed that their prized cabbages originated either from the sweat of Jupiter, shed while nervously attempting to explain away the rival pronouncements of a pair of opposing oracles, or from the tears of Lycurgus, king of the Edonians, unluckily apprehended by the wine god Dionysus in the shortsighted act of tearing up the sacred grapevines. While trussed and awaiting his unspeakable punishment, he cried, and his tears became cabbages. In Roman hands, the cabbage was a vegetable of status. Apicius listed at least five different ways of preparing it, variously accompanied by cumin seed, mint, coriander, raisins, wine, leeks, almond flour, and green olives, and the Emperor Claudius, the story goes, once convoked the Senate to vote on whether corned beef and cabbage was the best of all possible dinner dishes. (Thc senators, no fools, voted a unanimous yes.) Medicinally, cabbage-eating was said to prevent drunkenness, a recurring Roman problem, to cure colic and paralysis, and to protect the eater from plague.

While trussed and awaiting his unspeakable punishment, Lycurgus cried, and his tears became cabbages.

The earliest of cultivated cabbages was almost certainly kale, *B. oleracea* var. *acephala*, a curly-leaved, nonheading cabbage akin to the southern favorite, collard greens. The thick leaves branch from a fibrous stalk: in Ireland, legend holds that the fairies ride kale stalks for horses on moonlit nights. More expensive kale leaves, carved in green jade, paved the sarcophagus of the Pharoah Akhenaton's otherwise ransacked pyramid. Kale arrived early in colonial America, and was firmly established by the late eighteenth century, when an advertisement for "Dutch kale of various Colours; Scotch ditto" appeared in the *Maryland Journal and Baltimore Daily Advertiser*. "Scotch ditto" seems an off-hand tribute to the top kale consumers of Europe. Scotland's kale broth is one of the few survivors of kale vegetable cuisine, a school of cookery that admittedly was never extensive, even at its peak. Historically, kale has been a more popular food for livestock than for people, though for both it is notably nutritious, rich in calcium and vitamins A and C. Ornamental, or flowering, kales, which no one is actually expected to eat, have done somewhat better. "Horribly colorful" kales in reds and purples were popular centerpieces for Victorian dinner tables.

The Romans may have also had kohlrabi. Pliny, in the first century A.D., describes "a Brassica in which the stem is thin just above the roots, but swells out in the region that bears the leaves, which are few and slender"—which certainly sounds like kohlrabi, though one latter-day interpreter feels he was talking about cauliflower. Kohlrabi is sometimes called turnip-rooted cabbage, because of its turnip-like shape, but the more anatomically correct nickname is stem-cabbage, since the edible portion is actually an above-ground swelling of the stem, modified for starch storage. The official scientific name is the tongue-twisting *B. oleracea* var. *gongylodes* sub-var. *caulorapa*. Charlemagne ordered it planted in his domains, but fed it only to cattle, under the urgings of the royal physician, who warned that it would turn his soldiers unaggressively bovine. Kohlrabi is fairly bland in flavor—Alice B. Toklas of the notorious brownies (actually "Hashish Fudge") ascribed to it "the pungency of a high-born radish bred to a low-brow cucumber"—and has traditionally been best liked on the dinner tables of Germany and eastern Europe. It never made much culinary headway in America, where nineteenth-century seed catalogs usually carried it with the proviso that if the family spurned it, it would do just fine as a livestock feed. Burpee, in 1888, offered three cultivars, green, white, and purple, and had little to say about any of them.

The Emperor Claudius, the story goes, once convoked the Senate to vote on whether corned beef and cabbage was the best of all possible dinner dishes. (The senators, no fools, voted a unanimous yes.)

Heading cabbage, of the sort that dominates New England boiled dinner, had appeared by the time Julius Caesar invaded Britain, since his troops brought it with them, in two colors, red and green. In *B. oleracea* var. *capitata* the leaves are larger and thicker than those of non-heading cabbage, and are tightly wrapped around the central bud, on a truncated stem. The scientific name comes from the Latin *caput*, "head," as in *capitalist* and *decapitate*; the common name *cabbage* similarly comes from the Old French word for head, *caboche*. Writings from the fourteenth and fifteenth centuries mention "Caboches" and "cabogis," and even more frequently "coleworts," from the German, where cultivated cabbages are found on record in the twelfth century.

By the late sixteenth century, John Gerard lists sixteen different kinds of cabbages in cultivation, and credits them with a host of medical virtues, including the ability to clear the complexion and cure deafness. Cabbage juice was recommended for dog bites (mad and non-mad), serpent bites, mushroom poisoning, and bruises; the seed was used, in soup, for colic and consumption. Gerard warns vineyard owners off them, however: cabbages were reputedly capable of killing grapevines. The early English cabbage is commemorated in stone, on the monument of Sir Anthony Ashley of Wimborne St. Giles, Dorset, who died in 1627. He lies with an impressive cabbage at his feet, though no one these days is sure just why; the legend that he was the first cabbage-planter in England is certainly wrong. One food historian has suggested that perhaps he made some useful improvement in the existing cabbage breeds.

If so, Ashley's improved cabbage may have eventually reached the American colonies—though not with the first lot of transatlantic cabbages, which were planted in Canada in the 1540s by Jacques Cartier. John Winthrop, Jr., brought along an ounce of "Cabedg seed" in his seed chest, and over the course of the next century cabbages, gamely multiplying, became a staple of the colonial garden. By the mid-eighteenth century, garden writer Batty Langley announced that "to make an Attempt of informing Mankind what a Cabbage, Savoy, or Colly-flower is, would be both a ridiculous and simple Thing, seeing that every Person living are perfectly acquainted therewith. . . ." Among those perfectly acquainted was the inevitable Thomas Jefferson, who grew twenty-two different kinds at Monticello. Frequently mentioned are drumhead cabbages, with flattened tops reminiscent of the percussion instrument, and the darkly ruffled Savoys, which Jefferson described in glowingly classical terms as having a "wrinkled jade green head with a surface like crackled faience." It was their resemblance to the Savoy that led flower-growers to dub the old-fashioned damask roses "Cabbage roses."

Cabbage juice was recommended for dog bites (mad and non-mad), serpent bites, mushroom poisoning, and bruises; the seed was used, in soup, for colic and consumption.

W. Atlee Burpee's seed company was essentially founded on a cabbage, the Surehead, introduced to American gardeners in 1877. Response was enthusiastic: R. McCrone of Sykesville, Maryland, said his Sureheads were as large as water buckets; Seth Fish of Monmouth, Maine, called them "the finest sight on my farm"; J. M. Carroll of Springville, Alabama, said his gave entire satsifaction; and S. C. Stratton of Leetonia, Ohio, said his weighed thirty-five pounds apiece. By 1888, Burpee was offering thirty-one cabbage cultivars in addition to the Surehead, as opposed to one each of broccoli (in purple) and brussels sprouts.

By the late nineteenth century, however, the culinary cabbage was at a social rock-bottom, generally considered a food of the poor and the vulgar. Its banishment from the best kitchens was due to cabbage b.o., the penetratingly unpleasant smell generated in cooking. The odoriferous culprits here are a series of sulfur-containing compounds, volatile by-products of mustard oil and isothiocyanates. When heated, these normal constituents of cabbage break down and release repellent vapors of hydrogen sulfide, which smells of rotten eggs, ammonia, miscellaneous mercaptans, and methyl sulfide. Mustard oils are present in the cabbage plant as defensive chemicals; in an uncooked state, they serve to repel cabbage-eating pests. Not, however, people, who find them tastily appealing. In undisturbed cabbage tissue, the oils are bound to sugar molecules, as mustard oil glycosides, and are peacefully benign. When the cells are broken, as by unwary nibbling insects, enzymes clip off the restraining sugars and release the burning oils. These oils, mild in cabbage, cauliflower, and broccoli, put the spice in hot-dog mustard and the burn in horseradish.

Along with the mustard oils, members of the cabbage family contain goitrins, capable of blocking iodine uptake by the thyroid gland, and antivitamins, which bind to or mimic the real McCoys, preventing their uptake by the body. Neither is present in particularly threatening concentrations in the cultivated cabbage plant, which has been defused by centuries of human selection; wild cabbages, however, can contain up to four times the amount of the mild house-and-garden varieties.

While the nineteenth-century cabbage was most commonly subjected to prolonged boiling, the cabbage-fancying Pennsylvania Germans also turned theirs into sauerkraut. Sauerkraut is an ancient food tradition: the coolies who built the fifteen-hundred-mile-long Great Wall of China were nourished on cabbage pickled in wine. Centuries later, Genghis Khan's cohorts added salt and took the portable result along on their invasion of eastern Europe. The concoction, happily adopted by the invadees, outlasted the Mongol hordes. Rich in vitamin C, it was carried as an antiscorbutic on early sea voyages. (The British Navy gave it a try before latching on to limes.) It was dubbed *sauerkraut* in Austria; an Oriental version persists as *kimchi*.

Basically, sauerkraut is a fermented pickle. It is made by adding salt to shredded fresh cabbage, such that the internal water is drawn out of the sliced leaves by osmosis, and then adding water until the future

Sauerkraut is an ancient food tradition: the coolies who built the fifteen-hundred-mile-long Great Wall of China were nourished on cabbage pickled in wine. Centuries later, Genghis Khan's cohorts added salt and took the portable result along on their invasion of eastern Europe.

Broccoli buds, left to themselves, will eventually open—hence "sprouting broccoli"—while the cauliflower head, a degenerate and sterile inflorescence, lacks this ability, and like Peter Pan, florally speaking, never grows up.

sauerkraut is completely covered. The taste of the submerged product is due to the efforts of a specific microorganism, the bacterium *Leuconosotoc mesenteroides*, whose growth is promoted at a salt concentration of 2.25 percent. In a temperature range of 65–70 degrees F, *L. mesenteroides* multiplies, consuming sugar and spewing out lactic acid and an array of minor flavor components, including carbon dioxide, acetic acid, ethanol, mannitol, dextran, and miscellaneous esters. When the lactic acid concentration of the sauerkraut brew reaches 1 percent, the growth of *L. mesenteroides* grinds to a halt and the process is taken over by a second bacterial population, the lactic-acid-producing *Lactobacillus plantarum* and *Lactobacillus brevis*. Provided *L. mesenteroides* has left them enough sugar to work with, this dynamic duo will continue to pump out lactic acid, which can reach final concentrations of up to 2.4 percent. Ideal, however, is a concentration of 1.7 percent acid, usually reached after a fermentation period of two to three weeks. A less scientific description of "Sour-Crout" making is recounted in *The Farmer's Everyday Book* (1850):

"Cabbage is sliced up fine, and a layer of it is placed in the bottom of a barrel, which is plentifully salted; it is then well bruised with a heavy mall or pestle, or is trodden down by a pair of heavy boots, till the barrel is half filled with the froth that arises from the operation. Successive layers of cabbage and salt are added in this manner, each receiving the same treatment, till the vessel is nearly full. Some cold water is then poured in, and the top of the barrel is pressed down with heavy stones. The contents undergo a brisk fermentation, which continues a week or two, during which time the brine must be drawn off, and replaced by new, until it remains perfectly clear, when the process is finished."

Sauerkraut was said (by royalty-courting Pennsylvanians) to be a favorite food of Charlemagne; and President James Buchanan, a Lancaster boy, was famed for the sauerkraut suppers served at his home estate, Wheatlands. During World War I, nobody gave it up, but ate it under the patriotic sobriquet "Liberty Cabbage," and to this day in Maryland, sauerkraut is a standard accompaniment to the all-American Thanksgiving turkey.

The Pennsylvanians sliced up their barrels' worth

of cabbages with a *kohlhobel*, or cabbage plane, a device admiringly described by naturalist Peter Kalm during his late-eighteenth-century travels through North America: "This method of slicing cabbage is much more efficient than with ordinary knives: A tray was made of boards with a flat even bottom about three feet long and seven inches wide and with two-inch sides. In the middle of the tray was a large square opening about four inches wide. Across this were placed three knives parallel to one another. The width of each knife was one and a half inches. The edges were set aslant in a plane. The cabbage was grated by these knives." Kalm also had a good word for cole slaw, a sliced cabbage salad served up for supper with a dressing of vinegar.

Both broccoli and cauliflower are edible modifications of the cabbage flower. Botanists hypothesize that one of these is an early developmental form of the other, though considerable confusion exists over which is which. The edible portions of the broccoli plant include both fleshy stalks and clustered flower buds. These buds, left to themselves, will eventually open—hence "sprouting broccoli"—while the cauliflower head, a degenerate and sterile inflorescence, lacks this ability, and like Peter Pan, florally speaking, never grows up. This suggests, some scientists feel, that broccoli appeared on the scene first. Today most broccolis are green, but common nineteenth-century garden varieties were more colorful, in purples, browns, reds, and creams. Broccoli, *B. oleracea* var. *italica*, was apparently known to the Romans. Our common name is derived from the Latin *bracchium*, meaning branch, though Roman growers themselves referred to the plant poetically as "the five green fingers of Jupiter." Drusus, oldest son and heir of the Emperor Tiberius, is said to have been positively addicted to it. One story holds that the broccoli-besotted prince once refused to eat anything else for an entire month, at which point his urine turned a bright broccoli-green and his annoyed father, reportedly upbraiding him for "living precariously," ordered him to cease and desist.

Broccoli was introduced to France in the mid-1500s by the Italian Catherine dé Medicis (she also presented the French court with its first forks and sherbets) and soon after arrived in England, where it appeared on Elizabethan menus as "Brawcle." Phillip Miller mentioned it as "Italian asparagus" in his *Gardener's Dictionary* of 1724, and John Randolph described it in 1775 in his anonymously published *Treatise on Gardening*. "The stems will eat like Asparagus," said Randolph, "and the heads like Cauliflower." Still, broccoli never really caught on in the United States until the 1930s—and even then there were suspicious pockets of resistance. A famous *New Yorker* cartoon of the early broccoli period, captioned by E. B. White, shows a vegetable-proffering mother and uncooperative child:

"It's broccoli, dear."
"I say it's spinach and I say the hell with it."

Cauliflower, *B. oleracea* var. *botrytis*, seems to have been somewhat better accepted, though censured by Mark Twain as "nothing but a cabbage with a college education." This most intellectual of cabbages is a precocious annual, whose edible buds, in lieu of opening, solidify into a tightly packed head technically called a curd. Perhaps the earliest description of the cauliflower dates from the twelfth century, when an Arab botanist spoke of a "flowering Syrian cabbage." It was mentioned by John Gerard as "Cole flowery," and commented upon in seventeeth-century French documents as "chou-fleur." In Spain, according to Bert Greene, cauliflowers were grown ornamentally and it was customary for young women to wear small samples titillatingly positioned in their cleavages.

More commonly, however, the cauliflower was eaten, raw, boiled, or pickled, and by the eighteenth century, in one of these manifestations, it was a highly regarded addition to the dinner table. "Of all the flowers in the garden," announced Samuel Johnson, the way to whose heart was through the stomach, "I like the cauliflower." In France, the cauliflower was popularly eaten in soup—at court as the delectable Crème du Barry, a cream of cauliflower soup named for the equally delectable mistress of Louis XV. Most cauliflower cultivars nowadays are white or cream-colored—

the curds were once routinely blanched by wrapping them in the outer leaves to prevent unwanted discoloration by chlorophyll-inducing sunshine. Eighteenth-century gardeners also grew red and purple cauliflowers, which attractive plants seem to have gone out of fashion by the late 1800s. Burpee's 1888 *Farm Annual* offers eleven cauliflower cultivars, all white.

Latest of the inventive cabbages are the brussels sprouts, *B. oleracea* var. *gemmifera*, which serendipitously burst on the scene in mid-eighteenth-century Belgium. Earlier possible descriptions exist: some sources hold that the Romans had brussels sprouts, known as *bullata gemmifera*, or diamond-makers, from their putative ability to enhance mental prowess. (Mark Antony ate them, unsuccessfully, before meeting Caesar at the Battle of Actium.) Others suggest that the Roman references are really describing a very small form of heading cabbage. The sprouts are actually axillary buds, compact miniature cabbage heads sprouting from the stalk, which is topped by a rosette of large loose leaves. From these multiple buds comes the nickname "thousand-headed cabbage."

Brussels sprouts have been distinguished by their appearance in "Sprouts Gratin," the recipe that launched a million-dollar lawsuit in 1983–84. The complaint was plagiarism, an act difficult to prove in the cookery world, where recipes are often treated as simply up for grabs. The Sprouts Case, however, was definitively settled in favor of Richard Olney, author of *Simple French Food* (1974) in whose pages the fatal dish first appears.

Pricey as thousand-headed cabbages have proven to be, the world's most expensive cabbage patch to date is that of Xavier Roberts of Cleveland, Georgia, whose squashy-faced Cabbage Patch dolls, each with birth certificate and adoption papers, went on the market to multimillion-dollar sales in 1983. The original Kids, handmade at the Original Appalachian Artworks, sold for about a thousand dollars apiece; mass-produced models, though much cheaper, still cost a lot more than plain cabbages. Roberts, a sculptor whose innovative dolls are based on southern folk art, has had to deal with the protests of an organization of adopted children who feel it is demeaning to pretend that adopted infants are found in cabbage patches.

Cauliflower, *B. oleracea* var. *botrytis*, seems to have been somewhat better accepted, though censured by Mark Twain as "nothing but a cabbage with a college education."

TURNIPS

The turnip's finest hour occurred at dinnertime in the sixteenth century, when intricately carved in the shape of a turreted castle, a steepled cathedral, or a galleon (with lettuce sails), it served as the focal point of the "Grand Sallet" on the grandest of aristocratic tables. The turnip since has fallen steadily from horticultural favor. Fewer and fewer gardeners these days bother to grow them, and diners, presented with them, have a tendency to push them fretfully about on their plates and finally hide them under the mashed potatoes.

Plant scientists hypothesize dual centers of origin for the now-scorned turnip, one in the Mediterranean region, the other in Afghanistan and Pakistan. Oldest of the cultivated turnips is thought to be *Brassica rapa* var. *oleifera*, grown first in Asia for its oil-bearing seeds. Colza oil, pressed from the seeds of Indian turnips, was imported to fill the lamps of Europe from the thirteenth century onward. In Europe, however, the edible root crop seems to have preceded the oil-seed turnip. The Romans were familiar with it, ate it, and were said to prefer it to the carrot. Apicius' cookbook includes an upper-class recipe for turnips seasoned with myrtle berries and pickled in honey and vinegar, but the Roman turnip was more usually considered a food for the poor. It was similarly a cottage, rather than a castle, vegetable during the Middle Ages; in England, turnips appeared on the occasional family coat-of-arms to indicate a benefactor of the poor. Medicinally, the turnip, mashed and mixed with suet, was recommended for winter maladies: frozen feet, chilblains, and aching joints. It was also used to treat "goute," smallpox, and measles, and to make a nice "sope" for "beautyfying the face," a custom that may have been handed down from Roman times. Apicius directed wrinkle-conscious women of middle age to use facial masks of cooked turnip, cream, and mashed rosebuds.

Less frivolously, professional sixteenth-century cutlers used a mixture of turnip juice and earthworms to quench newly made knives hot from the forge, and individuals with nothing better to drink fermented turnips and turned them into alcohol.

The popularity of the turnip peaked in the seventeenth and eighteenth centuries, a period sometimes called the Age of the Turnip, when a fast-growing variety called the stubble turnip was introduced to European farmers. The respected Jethro Tull (1674–1741), inventor of the seed drill, is usually credited with suggesting that turnips would make a dandy winter fodder for livestock. A mix of chopped turnips was thus touted by progressive agriculturalists as "a Winter Pabulum for Cattle," and turnip fields were soon so extensive that Daniel Defoe, still ruminating over *Robinson Crusoe*, remarked on them during his tour of England in 1722. Turnips were grown in the American colonies from day one; in 1622, in Newbury, Massachusetts, a bushel of turnips would buy a cord of oak firewood. White and purple "Turneps" appeared on John Randolph's list of essential garden vegetables, and the enthusiastic Thomas Jefferson grew ten kinds.

By the nineteenth century, the gardener's ideal was the gigantic turnip. Thirty-pounders were standard fare, and a grower in California, home of the botanically sensational, reported harvesting a massive one hundred-pound root in 1850.

By the nineteenth century, the gardener's ideal was the gigantic turnip. Thirty-pounders were standard fare, and a grower in California, home of the botanically sensational, reported harvesting a massive one hundred-pound root in 1850. Turnip-heavy contemporary seed catalogs often offered up to twenty-five varieties. Burpee, in 1888, carried sixteen, including the White Egg; the Cowhorn, a white carrot-shaped variety; and the Golden Ball or Orange Jelly, noted for "rich, sweet, pulpy flesh."

The notable success of the turnip paved the way for other root crops. The rutabaga arrived in England in 1755, from Holland. A robust version of the turnip, the rutabaga, or "turnip-rooted cabbage," was developed by the Swiss botanist Gaspard Bauhin from a series of judicious turnip-cabbage crosses. The result turned out to be tailormade for the fields of the chilly north. It was most commonly eaten in Sweden, thus acquiring the nickname "Swedish turnip" or simply "swede." The official scientific designation is *Brassica napobrassica*.

The edible bottoms of both turnip and rutabaga are a combination of lower stem and taproot. Turnips thus develop partially in the ground and partially in the air. The lower two-thirds, all root, lurks underground, while the upper third, derived from the stem base, remains above. The above-ground portion, exposed to sunlight, accumulates an assortment of purple and red anthocyanin compounds and becomes pigmented, while the shielded root remains pale: hence the familiar turnip's purple top. The term *turnip-pate*, in common use in the seventeenth century, took into account only the snowy nether region. It was applied to individuals with very fair hair, those today called towheads or even platinum blondes. Both turnip and rutabaga are also biennials, the starch-laden hypocotyl and root botanically targeted for the nourishment of subsequent flower and seeds. Turnips are occasionally known to jump the gun and flower the first year, a circumstance once viewed with alarm. John Gerard, commenting ominously upon it in 1633, said, "the Turneps that floure the same year that they are sowen, are a degenerat kind, called Madneps, of their evill qualitie in causing frensie and giddinesse of the brayne for a season."

Many seed catalogs, slavishly following the trend, now leap insouciantly from tomatoes to watermelons without a turnipward glance.

The turnip, like cabbage, kohlrabi, kale, brussels sprouts, cauliflower, and broccoli, possesses, from a sexual point of view, perfect flowers, which are usually crosspollinated by proverbially busy bees. In many western European languages, however, the turnip, like many other ostensibly sexless objects, possesses gender; masculine in French, (*navet*), masculine in Spanish (*nabo*), feminine in German (*Rube*)—a linguistic curiosity that once led the touring Mark Twain to remark: "In German, a young lady has no sex, while a turnip has. Think what overwrought reverence that shows for the turnip, and what callous disrespect for the girl."

There's less overwrought reverence for the turnip nowadays—it doesn't even appear on the National Gardening Association's list of most popular American garden vegetables—which goes to show that the eating public is a notoriously fickle bunch. Many seed catalogs, slavishly following the trend, now leap insouciantly from tomatoes to watermelons without a turnipward glance. It seems a sad comedown for the root that ruled the vegetable roost for nearly two centuries, but fame, the poets tell us, is distressingly fleeting. Francis Bacon once wrote "Fame is like a river, that beareth up things light and swollen, and drowns things weighty and solid." It seems a fitting eiptaph for the turnip.

BEETS

The garden beet, *Beta vulgaris*, is said to get its formal name from the Greek letter *beta*, because the swollen root more or less resembles a Greek *B*. Less is more like it. As any casual student of the Greek alphabet can plainly see, the beet looks more like a sigma or even an omicron, than a beta. However, Linnaeus swallowed it, so there, taxonomically, matters rest. The beet is a member of Chenopodiaceae, the goosefoot family, so-called because the leaves of some of the more prominent members resemble the flat flappy feet of geese. Beet relatives include the pot-herbs Good-King-Henry and lamb's quarters, leafy garden spinach, and quinoa, a chenopod used by natives of the high Andes to make porridge and beer.

Botanists list three major types of beets: *B. vulgaris* ssp. *vulgaris*, which includes the familiar fat red beetroot, the sugar beet, and the out-sized mangel-wurzel; *B. vulgaris* ssp. *cicla*, the leaf beets or chards; and *B. vulgaris* ssp. *maritima*, the uncivilized sea-beet, believed to be the ancestor of all modern domesticated varieties. The oldest of cultivated beets were probably the chards, variously called the white-rooted beet, the silver beet, the spinach beet, and perpetual spinach. There is also a green-leafed variety with red stems and midribs known as the rhubarb beet. The most usual common name—chard— comes from the Latin *carduus*, meaning thistle. Chards, which have nothing remarkable in the way of root, are eaten much like spinach, for the leaves. One modern garden writer describes it as highly preferable to spinach, since it lacks spinach's "mouthdrying irony taste." Chard was domesticated by at least 2000 B.C. The Greeks and Romans grew it, in ruby-red and two shades of green, and cooks of the Middle Ages used it to make chard-and-sorrel soup.

The ancients, with remarkable lack of foresight, used the beetroot only medicinally. Pliny, who ate his beets as leaves, spoke slightingly of the "crimson nether parts" as the preserve of doctors and druggists, and the Anglo-Saxons, who got their beets from the Romans, used the pounded roots to make a bone salve. Beetroot, white or red, seems to have arrived on the vegetable scene in Italy around the second or third century A.D., developed by the process of selection from a scrawny wild ancestor. The Romans distributed their new acquisition throughout Europe, where it was known as late as the sixteenth century as the "Roman beet." The sixteenth-century beet was boiled in stews, puréed and baked in tarts, and occasionally roasted whole in the embers—all sounding tasty enough, though the Elizabethans may not have reaped full gourmet benefit from their beets, since it was recommended that the vegetable be wiped with fresh dung prior to cooking. Still, John Gerard described the red beet in 1597 as "not only pleasant to the taste but also delightful to the eie," which argues that a few cooks must have ignored the dung injunction. White beets seem to have been more common, and consequently less desirable, than red. In the 1570s, Thomas Hill, a gardener much given to imaginative but impractical advice, addressed himself to unhappy white beet growers: "To have the Beete growe redde, water the plant with redde Wine Lees." Hill also included his personal technique for producing bigger beets, by placing a "broad Tile, potshearde, or some other thing of weight" on top of the developing stalk, which naturally makes one wonder about the state of the Hill family beet patch.

A TART OF BEET ROOTS

From Giles Rose, Master Cook to Charles II, 1682

◆ First roste your Beet Roots in the Embers, and peel them very well, cut them in pieces, and give them a boil with a Glass of white-wine, and then beat them in a Morter, with a piece of Sugar, a little Salt, and Cinnamon, and put them into fine Paste with some green Citron rasp'd, and a piece of Butter, and do not cover it, but when it is baked serve it away.

The beet, though relatively impervious to red wine lees, is made red by anthocyanin, the same complex ring molecules that put the red in cabbages, apples, roses, and opium poppies. Predominant among beet pigments is the bright-red betacyanin, the compound that led nineteenth-century belles to use beet juice as rouge. Not everyone is able to metabolize betacyanin, an ability that is governed by a single genetic locus. Individuals carrying two recessive genes at the crucial point simply pass betacyanin, unharmed and intact, through their systems, which means that after a beet orgy, they pee pink.

The American colonists toted beets across the Atlantic, where both chard and beetroot, in red, white, and yellow, were well established by the eighteenth century. The opinionated Amelia Simmons mentions them in *American Cookery*, casting in her lot with the red: "The *red* is the richest and best approved; the white has a sickish sweetness, which is disliked by many."

The sickish white beet was viewed more positively by European chemists who, in the eighteenth century, were in the process of developing the sugar beet. In 1747 a Prussian scientist named Andreas Marggraf analyzed beetroots and found that they contained sucrose, the same profitably sweet ingredient contained in sugar cane. Marggraf's student Franz Archard instituted a program to improve the sugar potential of the existing beet crop and devised a process for efficiently extracting sugar from beets—a scientific accomplishment that overexcited him to the point of promising future outflows of tobacco, molasses, rum, coffee, vinegar, and beer, all from the miraculous beet. The king of Prussia, who took much of this with a grain of salt, still offered Achard a subsidy to establish the sugar beet industry, and the first processing plant was constructed at Kunern, Silesia, in 1801. The beets processed were Silesian fodder beets, which contained up to 6.2 percent sugar, as opposed to the 2 percent sugar content of the run-of-the-mill garden beet. Over the next decade, the sugar beet industry soared in importance, notably in France, where Napoleonic warmongering and the resultant British blockade of French seaports had cut off the French supply of West Indian cane sugar. In the ensuing scramble for sweeteners, factories manufactured syrups from raisins and honey—distressingly unsatisfactory substitutes for table sugar—until a pharmacist named Nicolas Deyeux, who seems to have kept current with the scientific literature, proposed using Achard's sugar beet. The first imperial sugar beet factory was established in 1812, three years before the imperial government went under at the Battle of Waterloo.

Today, sugar cane and sugar beets account for most of the world's sugar, with a little help from corn and sorghum, sugar maples and honeybees. The modern improved sugar beet, *B. vulgaris* var. *saccharifera*,

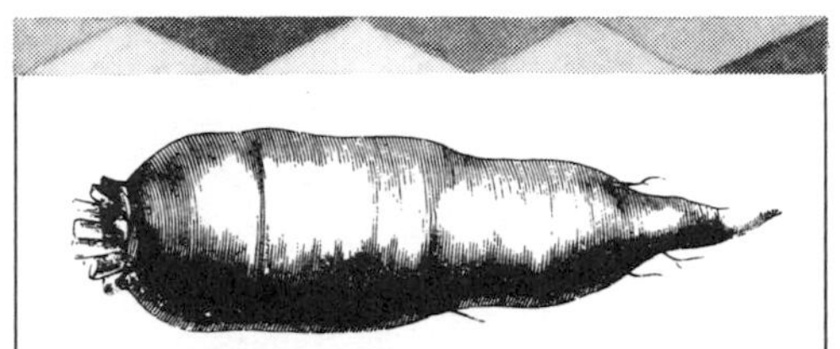

The sixteenth-century beet was boiled in stews, puréed and baked in tarts, and occasionally roasted whole in the embers—all sounding tasty enough, though the Elizabethans may not have reaped full gourmet benefit from their beets, since it was recommended that the vegetable be wiped with fresh dung prior to cooking.

contains an impressive 20 percent sucrose by weight and is now in fifth place on the FAO list of the world's major crops. (Number one: sugar cane.)

The eighteenth century also saw the introduction of the mangel-wurzel, like the sugar beet an offshoot of the early fodder beet. It was developed in Germany and Holland as a livestock feed and introduced to England in the 1770s, where an unfortunate mistranslation of the German *mangold-wurzel* ("beet-root") as *mangel-wurzel* ("scarcity-root") fostered the belief that these beets would make a dandy food for the poor in periods of famine. They were better suited to cows. Martha Washington experimented with them at Mount Vernon, and by 1888 Burpee's *Farm Annual* offered seven different types of mangels, including the Golden Tankard, which in illustration looks like a gallon jug with a ridiculously undersized topknot of leaves. Cows fed on it, says Mr. Burpee, give higher-priced milk. The *Farm Annual* also offers twelve varieties of garden beets, and a lone cultivar of chard, the stems and midribs of which were either cooked "like Asparagus" or pickled. Beetroot leaves—beet greens—are also traditional favorites, often collected, cooked, and served in a form preceded by the phrase "a mess of": a cup-sized mess contains 7,400 mg of vitamin A. Beet bottoms are often pickled ("horribly drowned in vinegar," says one offended beet lover), or converted more elaborately into borscht or red-flannel hash. In England, a couple of red beets are sometimes thrown in with the apples during cider-pressing to give the finished product a rich golden color, and a small beet in the stewing apples, says Bert Greene, turns out an attractively pale-pink applesauce.

Over the next decade, the sugar beet industry soared in importance, notably in France, where Napoleonic warmongering and the resultant British blockade of French seaports had cut off the French supply of West Indian cane sugar.

Embarrassed persons have been known to turn red as a beet, or the carnivorous equivalent, red as a turkey-cock, since colonial times. The former red, exactly like a blush, is impermanent. Used as fabric dye, beets make an initial deep rosy-red that, disappointingly, promptly comes out in the wash. Better are red beans, which natural dyers use to produce a rich red-brown, and best is madder, *Rubia tinctoria*, a sprawling perennial of the same plant family as coffee and quinine. The British redcoats were dyed with madder root, as were the beet-red stripes in Betsy Ross's famous flag.

SPINACH

Popeye, according to the latest in athletic-oriented nutrition, would have done better for himself with a plate of spaghetti. *Spinacia oleracea*, garden spinach, for all its vaunted muscle-building power, doesn't contain much of anything but vitamin A—of which, granted, it contains a lot: 14,500 units per cooked cupful. Vitamin A is great for night vision, which means that Popeye the Sailorman would have had an edge as a night pilot, but is less of a boost for dockside brawls. Still, spinach has its own niche in the vegetable world. The leaves, cooked and raw, have been on the human menu for over a thousand years; and the juice—deeply and intensely green—was used as a food coloring up through the nineteenth century, and, seasonally, to make touchpaper for fireworks. Soaked in spinach juice and dried, paper smouldered well, just right for touching off Roman candles and Catherine wheels.

Like fireworks, spinach comes from Asia, though botanists believe it originated closer to Persia than to China, and the modern name derives from the Persian *isfānākh*, which means green hand. The Persians, who initially cultivated spinach for the delectation of their exotically long-haired cats, sent the vegetable east toward China well before it arrived in the West. Chinese records, which refer to it as the "Persian herb," note its arrival in A.D. 647 as a gift from the intermediately located king of Nepal. The Chinese liked it and planted it around the edges of their vast rice paddies, where it flourished. It took another four hundred years for the Persian herb to reach Europe, where it showed up, via the conquering Moors, in eleventh-century Spain. Mention survives of a treatise on spinach written by Ibn-Had-Jadj, a Spanish Moor, who addressed it enthusiastically as "the prince of vegetables." It's uncertain when the princely leaf reached northern Europe: "Spinachium" appears on a 1351 list of the unexciting fare permitted monks on fast days, and "spynoches" can be found in *The Forme of Cury*, a recipe collection written around 1390 by the master chefs of Richard II, whom they described in their apple-polishing introduction as "the best and ryallest viander of all Christian kings." Europeans seem to have been of two minds about spinach. John Gerard, anticipating generations of American children, said it was watery and tasteless, but Catherine dé Medicis was reportedly so mad for it that to this day the phrase *à la florentine* attached to French dishes means with spinach, in honor of Catherine's Italian hometown of Florence.

Catherine dé Medicis was reportedly so mad for it that to this day the phrase *à la florentine* attached to French dishes means with spinach, in honor of Catherine's Italian hometown of Florence.

Perhaps it was all in the preparation. The ancient Medes recommended washing each spinach leaf twelve times before dropping it into the cooking pot: eleven times in clear water, for best results, with a final rinse in human tears. "The dayly eating of the hearbe Spinage doth marvellously profite such having a horse voice, and that hardly fetch breath," wrote an English spinach fan, "if the hearbe after proper seething and ordering be either fried with sweete butter or the oyle of Almonds, and that to it Pepper bruised be wittily added." Another claimed that spinach chopped in oatmeal made a "sublime Pottage," and the Dutch recommended it baked in tarts. The French compared it poetically to virgin beeswax, since it could adapt innocuously to any culinary situation. It was thus said to require a great chef to do justice to its delicate flavor. Apparently such were around, as Louis XIV, deprived of spinach by a conscientious royal physician, is said to have bellowed in chagrin, "What! I am king of France and I cannot eat spinach?" He could and did.

Spinach arrived in North America at least by the early seventeenth century, as "1 oz. spynadg" appeared in John Winthrop, Jr.'s well-documented seed list. Bert Greene suggests that an early Puritanical children's prayer begging divine protection from fire, famine, flood, and "unclean foreign leaves," refers in the last instance to spinach. Tobacco seems a more likely guess; anything as undeniably healthy and inherently repulsive to children as spinach surely met with wholehearted Puritan approval.

Spinach, though grown, was not a major obsession of colonial gardeners. Bernard M'Mahon in 1806 mentions only three existing cultivars. The first major attention paid spinach in this country seems to have come from David Landreth, founder of the D. Landreth Seed Company, established in Philadelphia in 1784. The Landreth Seed Company is still going strong; now based in Baltimore, it is the oldest continuing seed dealership in the United States. Among its claims to fame are the development of the first truly white variety of potato, the introduction of the tomato to garden cultivation, and the first American listing of cantaloupe seed, imported from a source in Tripoli. Landreth, who counted Washington, Adams, Jefferson, and Monroe among his early customers (and once patriotically granted Washington a thirty-day extension on payment of his bill), developed a notable slow-bolting spinach called Bloomsdale after the company farm in Bristol, Pennsylvania. The improved offspring of Bloomsdale spinach are still around and still popular today, glossily dark green with thick crumpled leaves.

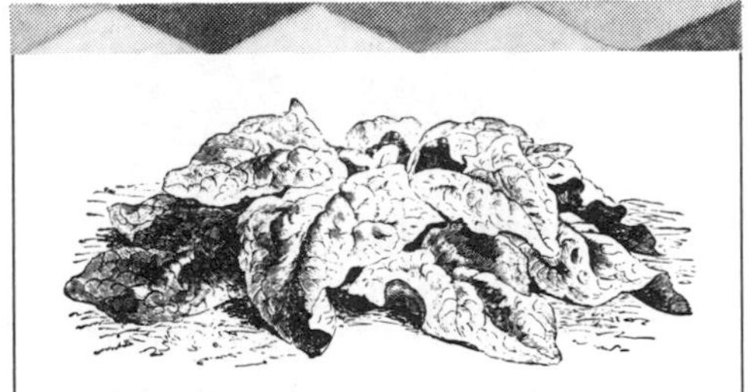

The ancient Medes recommended washing each spinach leaf twelve times before dropping it into the cooking pot: eleven times in clear water, for best results, with a final rinse in human tears.

Bloomsdale is a smooth-seeded spinach, the smooth seeds a feature that appeared, presumably by spontaneous mutation, at some point in the sixteenth century. The older and more primitive spinaches produced prickly seeds. Neither, in a strict botanical sense, is a seed at all, but a utricle (fruit) encased in a smooth or spiny capsule. The prickly seeded (fruited) varieties, nicknamed winter spinaches in the mistaken belief that they were more resistant to cold, have been traditionally more common in this country, though food critics deem them less flavorful than their smooth-seeded relations. Smooth-seeded "summer" spinach predominates in Europe.

The sexually normal spinach plant is dioecious, with approximately equal proportions of male and female plants. Versatile variations, however, appear on either side of the sexual average: thus, there are both perfect-flowered hermaphroditic spinaches, which consolidate all the necessary sexual equipment on one handy plant, and monoecious spinaches, which produce high proportions of all-female or all-male plants. All are annuals, and most—over 50 percent of the commercial crop in this country—are grown in California. "The Spinach Capitol of the World," however, at least according to residents, is farther east, in Zevala County, Texas. Chances are it isn't: at least, Gilroy, California, "The Garlic Capitol of the World," isn't, and East Grand Forks, Minnesota, "The Potato Capitol of the World," isn't, either. Hollister, California, "The Earthquake Capitol of the World," may actually have a shot at it: four major faults run through Hollister, and the town is a hotbed of budding seismologists.

Landreth, who counted Washington, Adams, Jefferson, and Monroe among his early customers (and once patriotically granted Washington a thirty-day extension on payment of his bill), developed a notable slow-bolting spinach called Bloomsdale after the company farm in Bristol, Pennsylvania.

Far from restricting itself chastely to *Spinacia oleracea*, the term *spinach* is frequently used in a generic sense to mean practically anything leafy and green. Captain Cook discovered New Zealand spinach, *Tetragonia expansa*, on his landmark voyage down under in 1771; though edible, it was initially grown back home in England as a houseplant, for its attractive fleshily triangular leaves. Good-King-Henry, *Chenopodium bonus-henricus*, is commonly known as wild spinach; orach, *Atriplex hortensis*, whose elegant pearly gray leaves were popular in colonial salads, is nicknamed mountain spinach. Chinese spinach, also called Joseph's coat for its many-colored leaves, is an amaranth.

"Spinach!" has meant nonsense since at least the 1920s, probably from the mid-nineteenth-century "gammon and spinach," as in Charles Dickens's rueful "What a world of gammon and spinage it is, though, ain't it!" Spinach thus fell in with such anti-bombast expressions as humbug, twaddle, baloney, moonshine, balderdash, hogwash, horsefeathers, and Go tell it to the Marines! —and when Will Rogers, in 1924, told the nation that all politics is applesauce, he could just as well have said spinach.

ONIONS

Since the 1930s, "to know one's onions" has meant to be well informed, on top of things, competent, equal to the odd emergency. Taken literally, it's a phrase easier said than done. The common onion, scientifically known as *Allium cepa*, is a member of a vast and varied genus classified, depending on the state of scientific debate, under either the lily family (Liliaceae) or the amaryllis family (Amaryllidaceae). Just now, according to the authoritative horticultural dictionary *Hortus Third*, the onion is an amaryllis. There are diehards, however, who still insist that it's a lily, and there's even a small botanically subversive group that assigns the onion to a separate family all its own (Alliceae). There are over six hundred species of alliums. As well as the common onion, edible alliums include shallots and potato onions (*A. Cepa aggregatum*), Egyptian or tree onions (*A. Cepa proliferum*), Welsh bunching onions (*A. fistulosum*), chives (*A. Schoenoprasum*), garlic chives (*A. tuberosum*), rakkyo (*A. chinense*), leeks, kurrats, and rocambole (*A. Ampeloprasum*), and garlic (*A. sativum*).

Unlike the ordinary onion, where the rule is one seed, one onion, shallots and potato onions are multipliers, and accordingly much more generous with their returns. Shallots are named for the ancient city of Ascalon (now Ashkelon) in Israel, where they were once intensively cultivated. They produce loose clusters of bulbs or cloves, milder-tasting than onions and, unless homegrown, much more expensive. Potato onions first arrived in this country in the early nineteenth century: New York seedsman Grant Thorburn offered them as a new introduction in 1828. Larger than shallots, these produce seven or eight deep-yellow-skinned lateral bulbs per plant. Their number and underground location apparently reminded some early grower of potatoes.

Egyptian onions, also called top, or tree, onions, were unknown in Egypt, but grow wild throughout temperate North America. These peculiar perennials bear their bulbs at the tips of the leaf stalks, hence "top" onion. Both bulbs and leaves are edible. Similarly perennial are chives and garlic-flavored Chinese chives, grown for their tangy leaf stalks, and Welsh bunching onions, which originated not in Wales, but in eastern Asia. The "Welsh" is believed to be a corruption of the German *Welsch*, meaning foreign. Leeks, sometimes called the poor man's asparagus, look at first glance like obese scallions. They do not form bulbs, but are grown for their enlarged leaf bases, as are the related Mediterranean kurrats. Rocambole, also called sand leek or serpent garlic, produces both underground bulbs and aboveground bulbils (edible) at the tips of twisted snake-like stalks. True garlic, multicloved and potent, is beloved of herbal medics and Italian cooks and anathema to vampires and cabbage worms.

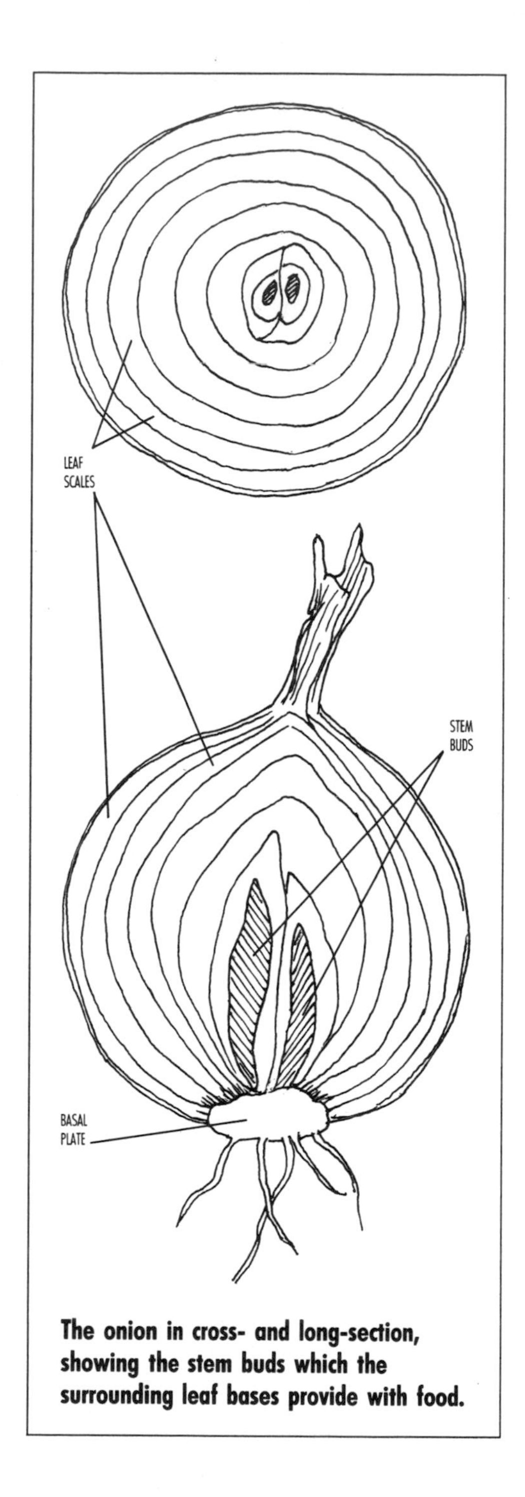

The onion in cross- and long-section, showing the stem buds which the surrounding leaf bases provide with food.

There are also numbers of highly ornamental alliums, or "flowering onions," among them *A. moly*, the lily leek or golden garlic, the plant that kept Circe from turning Ulysses into a pig, and the spectacular *A. giganteum*, which bears lavender flowers the size of small grapefruits on towering four-foot stems. Onion flowers are all amaryllis-like, a telling taxonomic point. The flower cluster (inflorescence) is umbellate, meaning that the stalk or stem terminates in one or more individual flowers all borne from a common point. The word comes from the Latin *umbella*, which means parasol, since such flowers are often parasol shaped.

In the common vegetable garden, the common onion rarely reaches the point of flowering. *A. cepa* is biennial: the tasty bulb that gardeners seasonally yank was intended by the onion plants as food for the following year's flowers and seeds. Onions, like tulips and tiger lilies, form true bulbs made up of a series of tightly overlapping fleshy leaf bases, or scales, surrounding a central bud. The dry crackly outer leaves, generally referred to as the skin and useful for dyeing Easter eggs, are formally and collectively known as the tunic. The scales, crammed with water and starch, are held together by a basal plate of stem tissue, from which roots will develop if the onion bulb is planted rather than eaten. Because the onion bulb is a single united entity rather than a conglomeration of separate cloves as found in garlic, it was referred to by the Romans as *unio*, meaning united. From *unio* came the medieval French *oignon*, the Anglo-Saxon *onyon*, and eventually the modern *onion*.

Onions were among the earliest of cultivated foods and probably among the first vegetables routinely nabbed by primitive hunter-gatherers, who could have easily identified them by their distinctive smell. They are believed to have originated in central Asia and have been domesticated since about 3000 B.C. Moslem legend imprecisely dates them to the exit from the Garden of Eden: As Satan hastily departed, the angel with the flaming sword hot on his heels, onions are said to have sprung from his right footprint and garlic from his left.

The earliest known written reference to the onion is a Sumerian cunieform tablet from about 2400 B.C., in which the onion appears as an innocent bystander in a complaint against the city governor, who had illegally co-opted the temple oxen to plow his onion and cucumber patches. It is more impressively immortalized in an inscription of the Great Pyramid at Giza, where it is recorded that during the construction period, sixteen thousand talents were spent on onions, radishes, and garlic to feed the laborers. A talent, worth some fifty-six pounds of solid silver, would have bought a good many onions—and all doubtless needed, since the Great Pyramid covers thirteen acres, contains two and a half million two-ton blocks of stone, and took one hundred thousand men over twenty years to build. Onions were popular among the ancient Egyptians in more ways than one; they were also used in the mummification process, stuffed into the body cavity along with sawdust and myrrh.

The Greeks and the Romans ate onions, though in both societies onions were generally viewed as fare for the lower classes. The historian Herodotus may have been the first to advocate the athlete's onion: he advised those in training for the Olympic races to eat two onions a day, one the size of a fist upon arising, one the

Because the onion bulb is a single united entity rather than a conglomeration of separate cloves as found in garlic, it was referred to by the Romans as *unio*, meaning united. From *unio* came the medieval French *oignon*, the Anglo-Saxon *onyon*, and eventually the modern *onion*.

size of a thumb before going to bed. In Nero's Rome, gladiators were massaged with onion juice before entering the arena in order to keep their bodies firm.

Nero himself, however, is noted not so much for onions as for leeks, which he consumed in such quantities to sweeten his singing voice that he was nicknamed Porrophagus ("the leek-eater"). The Roman military, equally fond of leeks, distributed them across Europe in the wake of their interminable conquests and were no doubt responsible for their highly successful introduction to the British Isles. Once across the Channel, the leek went on to grace Scotland's traditional cock-a-leekie soup, to feed Chaucer's Canterbury-bound Pilgrims, and to become the national symbol of Wales. The Welsh connection, the most commonly recounted legend tells us, originated at a Briton-vs.-Saxon battle of A.D. 640 during which the Welsh combatants wore leeks in their hats to distinguish themselves from the enemy. Ever since, patriotic Welsh have worn leeks on St. David's Day, either in memory of the victorious event or in honor of the patron saint of Wales. The Welsh, incidentally, are also famed for their sweet singing voices.

Archaeologists uncovered a basket of onions in the ruins of Pompeii, in the biggest and best of the town brothels.

Archaeologists uncovered a basket of onions in the ruins of Pompeii, in the biggest and best of the town brothels, appropriate, since onions, especially those eaten young and green, were said to "serve for no other thing but to provoke and stirre folke to the act of carnal copulation." Nonetheless, onions, along with beans and cabbages, were the prime vegetables of the Middle Ages. Chopped and mixed with violets, they comprised a favorite savory at the dinner table; plain, they were recommended for dog bites, cystitis, and the stings of "venomous worms." Mixed with honey and hen grease, they were said to remove "red and blue spots" from the skin; thrown, they were said to protect a new bride from the Evil Eye.

By Elizabethan times, uses of the onion had multiplied. Medicinal uses ranged from hemorrhoids to blisters, and Queen Elizabeth's surgeon, William Clowes, used onion juice to treat gunpowder burns. John Gerard, who must never have tried it, claimed that "the juice of an onion annointed upon a bald head in the sun bringeth the haire again very speedily," but warned that indulgence in cooked onions could bring on headaches and dimness of vision. Seventeenth-century salads consisted mostly of onions and mixed herbs, and the health-minded ate onions with honey for breakfast as a general-purpose tonic. The poor, who consumed the lion's share of the onions, ate them raw like apples or roasted like potatoes. John Evelyn summed up the onion attitudes of the era in his 1699 edition of *Acetaria, A Discourse of Sallets*: "The best [onions] are such as are brought us out of Spain, and some that have weighed eight Pounds. Choose therefore the large, sound, white, and thin skinned. Being eaten crude and alone with Oil, Vinegar, and Pepper, we use them in Sallet, not so hot as Garlic, nor at all so rank. In Italy they frequently make a Sallet of Scallions, Chives, and Chibols only seasoned with Oil and Pepper, and an honest laborious Country-man, with good Bread, Salt, and a little Parsley, will make a contented Meal with a roasted Onion." To dream of such a contented meal, however, was an ill omen, said to warn of impending domestic disaster.

Avoided by aristocrats because of its "crude and vulgar smell," the onion first appeared on the elite dinner table cleverly disguised in soups, side dishes, and sauces. It figured prominently in the sixteenth-century French *restaurants*, which originally were meat-based soups spruced up with onions and herbs. By the eighteenth century, the name had acquired its modern meaning, spreading from the soup to the establishment in which it was eaten. French onion soup is said to have been created by the de-throned King Stanislaus I of Poland, father-in-law of Louis XV, who had time on his hands during his necessarily prolonged sojourn at his daughter's court. (Stanislaus is also noted for travelling across Europe disguised as a coachman and for the invention of baba au rhum.) The culinary versatility of the onion is best illustrated, however, by the story of an eighteenth-century French caterer who, faced with customers and no entrée, served up a pair of old water-buffalo leather gloves, shredded and simmered with onions, mustard, and vinegar. The recipients reported them excellent.

The most expensive onion ever eaten was con-

The sailor, who afterward remarked only that he thought it remarkably insipid-tasting for an onion, had snacked on a *Semper Augustus* tulip bulb worth fifteen hundred dollars on the open market.

sumed in neither soup nor inventive stew, and in fact turned out not to be an onion at all. It was offhandedly eaten by a nameless sailor in the 1630s on board a ship transporting, among other items of cargo, a load of tulip bulbs. The bulbs were headed for the gardens of the filthy rich: Europe at the time was in the throes of tulipomania, a craze that sent the price of individual tulip bulbs, newly introduced from the seraglios of Turkey, to astronomical heights. The sailor, who afterward remarked only that he thought it remarkably insipid-tasting for an onion, had snacked on a *Semper Augustus* bulb worth fifteen hundred dollars on the open market.

During the Civil War, doctors in the Union army routinely used onion juice to clean gunshot wounds, and General Grant, deprived of it, sent a testy memo to the War Department: "I will not move my troops without onions."

Domesticated onions—the yellow kind, still most common today on supermarket counters—came to America belowdecks on the *Mayflower* and were planted in the first Pilgrim gardens. Wild onions, the new settlers found, had arrived in force well before them: over seventy species of *Allium* are indigenous to North America, among them wild garlic, ramp, prairie onion, and tree onion. Some or all of these are said to have saved the Jesuit explorer Père Marquette and company from starvation while en route from Green Bay to the site of modern Chicago in the 1670s. The name *Chicago*, aptly, comes from the Indian *CicagaWuni*, "Place of Wild Garlic." The American Indians gathered alliums, ate them, and used them medicinally to treat bee stings, coughs, and the common cold.

The globe onion was ubiquitous in later colonial gardens, and was a great favorite of George Washington, who referred to it besottedly as "the most favored food that grows." Colonial onions were eaten roasted, boiled, or pickled, and a promising, if somewhat vague, pickling recipe survives from Harriott Pinckney Horry's *Receipt Book* of 1770: it involves soaking the onions in brine in the sun for two days, then immersing them in "strong Vinegar with a good deal of spice." Onions were used to treat insomnia (two or three, raw, eaten daily), pneumonia, diabetes, and rheumatism in human beings, and mange in animals. Onion juice was considered an effective antiseptic well into the nineteenth century. During the Civil War, doctors in the Union army routinely used it to clean gunshot wounds, and General Grant, deprived of it, sent a testy memo to the War Department: "I will not move my troops without onions." They sent him three cartloads.

The onions he got may well have been from the Connecticut River Valley, a region long famous for onion production. Advertisements for "choice Connecticut onion seed" were appearing by the 1760s, and Connecticut onions formed an integral part of the booming West Indies molasses trade, which in turn went into New England rum. Many of the earliest and most successful American-named onion varieties were Connecticut-bred, among them the Red Wethersfield (ca. 1800), the Southport Yellow Globe (ca. 1835), and the Danvers Yellow (ca. 1850). Today onions have moved west. Present-day top producers are California and Texas, which have both done their bit to put the onion in fourth place among American vegetable crops.

Yellow, or golden-globe, onions are now available year-round. They're generally tawny-skinned and sharply pungent, though newer hybrids, notably the Vidalia onions of Georgia and the WallaWalla onions of Oregon, are extraordinarily sweet, enough so that connoisseurs chomp them raw without a quiver. Milder yet are the larger (and seasonal) Spanish and Red Italian onions, both now grown in the United States, and the flattish white Bermuda onions, which still come from Bermuda.

Opinions vary over which of the many sweet onions now available actually is the sweetest—there's a certain amount of competitive rhetoric, for example, between Georgian and Oregonian growers—but the answer may soon be a matter of scientific record. Research chemists at the U.S. Department of Agriculture have developed a "sweetness meter," an analytical device based on onion translucence. While not quite crystal clear, onions do transmit light: about 1 percent of the light beamed at an onion passes right through the bulb and comes out the other side. The precise amount of transmitted light is related to the amount of dry matter in the onion, which in turn is related to onion sweetness. So far those most interested in metering sweetness have been onion growers and breeders, who use the technique to screen promising seed bulbs for future propagation.

ONION PORRIDGE

From Charles Elmé Francatelli, The Cook's Guide, and Housekeeper's & Butler's Assistant, *1861*

◆ Take a Spanish onion as big as you can procure, peel and split it into quarters, and put these into a small stewpan with a pint of water, a pat of butter, and a little salt; boil gently for half an hour; add a pinch of pepper, and eat the porridge just before retiring to bed. This is also an excellent remedy for colds, and was imported to me by a jolly, warm-hearted Yorkshire farmer.

Onions come in a wide range of sizes, shapes, and colors. Bulbs may be flat, round pear-shaped, or elongated, as in the foot-long onions of Japan. Colors include white (*Sturtevant's Edible Plants of the World* lists four grades of onion whites; plain, dull, silvery, and pearly), yellow-green, copper, salmon-pink, blood-red, and purple. All are characterized by the pungent onion smell, an odor that caused the cowboys to refer to Geroge Washington's most favored food as "skunk eggs." The determinedly volatile oils that produce the onion odor enter the lungs where, exhaled, they create the notorious onion breath. Fastidious onion-eaters have fussed about onion odor for centuries: proposed remedies have included post-onion mouthfuls of parsley, celery tops, coffee beans, cardamom seeds, and cloves. Pliny, the Roman naturalist, swore by roasted beetroots. Nothing actually does much good, and habitual onion-eaters may as well resign themselves to paying a price for their pleasure.

The wives, after thinking it over, decided that they preferred their onions to their husbands, so they used magical ropes made of eagle down to float up into the sky, where they remain as the Pleiades, presumably eating onions to their hearts' content.

The molecules that bring about this social stigma are sulfur-containing compounds, derivatives of the common amino acid cysteine. The primary constituent of onion oil, which contains some ten volatile chemicals, is n-propylthiol; that of garlic—one of the three most disliked odors in America, according to the *San Francisco Chronicle*—is diallyl disulfide. According to Indian legend, it's n-propylthiol that we have to thank for the Pleiades. A group of seven young Indian wives, the story goes, were fond of eating onions, but their husbands, disliking the smell of onion breath, became angry and forbade the practice. The wives, after thinking it over, decided that they preferred their onions to their husbands, so they used magical ropes made of eagle down to float up into the sky, where they remain as the Pleiades, presumably eating onions to their hearts' content.

Along with the odoriferous sulfur-based volatiles, onions exude pungent fumes that make the eyes water. The tear-inducing compound, powerful enough, said Benjamin Franklin, to "make even heirs and widows weep," is formally called the lachrymator, from the Latin *lacrima*, "tear." Its chemical structure, dickered over in laboratories since the 1950s, was definitively identified in 1979 by Eric Block and Robert Penn at the University of Missouri as a specific conformation of propanethial-S-oxide. Propanethial-S-oxide is so volatile that it barely hangs around long enough to study and is accordingly difficult to isolate and purify. It very rapidly hydrolyzes in water, such as that present in the human eye, to produce sulfuric acid, which, in turn, irritates the eye and causes tearing. Onion peelers can avoid a bout of weeping by chilling the onions prior to applying the knife—low temperature reduces the volatility of the lachrymator—or by peeling under cold running water, which dissolves the lachrymator before it reaches the eyeball.

To even out the balance sheet, however, the onion is also the repository of a number of biological benefits. Avoided by Elizabethans, who liked their ladies plump, on the grounds that it encouraged weight loss, the onion at thirty-eight calories a bulb, is tailor-made for the struggling twentieth-century dieter. It also contains useful quantities of potattsium, phosphorus, and vitamin C, and the yellow varieties are good sources of vitamin D. More impressively, onions have the distinction of being the only plants so far known to contain prostaglandins—specifically prostaglandin A1, a fatty acid derivative that acts to lower blood pressure. Its presence in onions, however, at a concentration of one part per million, is more botanically interesting than potentially therapeutic, despite reports that onions are effective antihypertensives.

Onion-eating isn't a taste for everybody, which, according to at least one garden writer, is a crying shame. "In the onion," writes Charles Dudley Warner grandly, "is the hope of universal brotherhood. If all men will eat onions at all times, they will come into a universal sympathy," it's a praiseworthy aspiration.

In the meantime, boiled onion juice is said to make a dandy polish for gold-leaf picture frames.

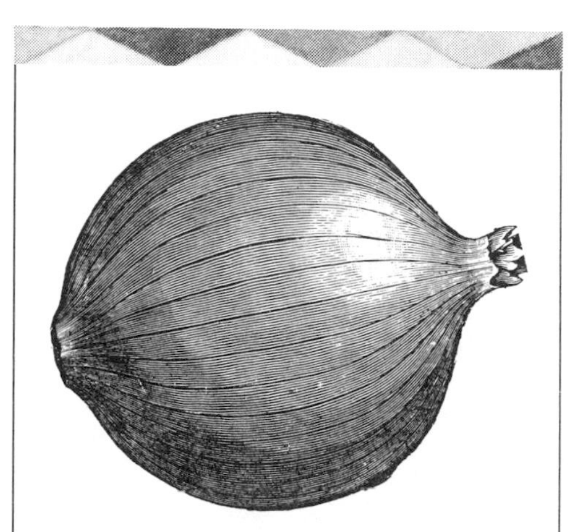

Onion peelers can avoid a bout of weeping by chilling the onions prior to applying the knife—low temperature reduces the volatility of the lachrymator—or by peeling under cold running water, which dissolves the lachrymator before it reaches the eyeball.

PUMPKINS AND SQUASHES

In 1699, Paul Dudley, a butterfingered citizen of Massachusetts, dropped a pumpkin in his pasture. The following year he harvested some 260 fruits of the disaster, an impressive enough haul to be recorded for posterity and a testimony to pumpkin survival skills. It's not recorded what Mr. Dudley did with his serendipitous pumpkins, but he doubtless put them to good use, since colonial New Englanders are said to have used more pumpkins in more ways than anyone else before or since. The prolific Dudley pumpkin and offspring were cucurbits, members of the large and nightmarishly complex Cucurbitaceae family. Prominent pumpkin relatives include squash, cucumbers, melons, and gourds, plus a few off-the-wall distant cousins, such as *Benincasa hispida*, the wax gourd, the waxy cuticle of which can be scraped off and used to make candles, and *Luffa cylindrica*, from whose dried fruits trendy bathers acquire their luffa sponges.

Pumpkins and squashes, so closely interconnected that many promiscuously interbreed to form misshapen, but usually edible, "squmpkins," have bewildered botanists for centuries. Common garden classification, which inconveniently bears little resemblance to accepted botanical order, divides the multitudinous squashes into summer and winter varieties. Summer squashes ripen in summer, have delicate edible shells and seeds, and should be eaten hot off the vine, since they have generally poor keeping qualities. Examples are the yellow crookneck, the bush scallop or pattypan, and the ubiquitous zucchini. Winter squashes ripen in the fall, have tough inedible shells and large hard seeds, and store well for periods of several months. Examples are the acorn, butternut, and Hubbard squashes, the more recently introduced spaghetti squash, and the orange topknotted turban squash.

Pumpkins, botanically lumped with the summer squashes, behave persistently like winter squashes, and many early naturalists and travellers seem to have used the name simply to indicate any fruit inordinately big and round. As late as 1885, the French seedsmen Vilmorin-Andrieux listed a long string of vegetable behemoths in *The Vegetable Garden* under the heading "Pumpkins," stating, "Under this name, which does not correspond to any botanical division, are grouped a certain number of varieties of *Cucurbita maxima* which are remarkable for the great size of their fruit." Included are the mammoth pumpkin; the Hubbard squash; the Valparaiso squash, shaped like a mammoth lemon; the chestnut squash, round and brick-red; and the turban gourd.

Pumpkins, botanically lumped with the summer squashes, behave persistently like winter squashes, and many early naturalists and travellers seem to have used the name simply to indicate any fruit inordinately big and round.

More discriminating taxonomists these days sort the edible squashes into four basic species. *Cucurbita pepo*, noted for pentagonal stems with prickly spines, encompasses all the summer squashes, pumpkins, acorn squashes, spaghetti squashes, and miscellaneous gourds. *C. maxima* (round stems) includes the banana, buttercup, Hubbard, mammoth, and turban squashes. *C. moschata* (pentagonal smooth stems) contains the butternut and the cushaw; *C. mixta*, the white and green cushaws and the Tennessee Sweet Potato squash. Amid all this variety—squashes come in an immense array of bizarre shapes and in colors ranging from tan, cream, and orange to blue, black, and salmon-pink— pumpkins continue to distinguish themselves in the matter of sheer size. Joshua Hempsted, an eighteenth-century Connecticut colonist, noted in his diary for 1721: "Wedensd. 20th: saw a pumpkin 5 foot 11 inches Round." Unfortunately, he didn't weigh it. A latter-day record-setter of 1983, also from Connecticut, weighed a mind-boggling 580 pounds—due, according to the grower, to injections of milk in the stem—and a variety called Atlantic Giant, developed in Nova Scotia for gigantism, now routinely produces fruit topping 400 pounds. Its top weight to date is 612 pounds, which would make, at a guess, a good three hundred pumpkin pies.

Average pumpkins nowadays run a more manageable ten to twenty pounds.

The very earliest cucurbits—probably originating in Central America—were considerably smaller. They were also unpleasantly bitter-fleshed, and are thought to have been valued by primitive man for their protein- and oil-rich seeds. Squash and pumpkin cultivation dates back nine thousand years, judging by scattered remains of seeds and stems found in prehistoric caves in the Tamaulipas mountains of Mexico. This makes them the first of the domesticated foods of the "Indian triad"—squash, beans, and maize—that formed the basis of pre-Columbian Indian diet in both North and South America. By the arrival of the Europeans, selection had produced squashes sizeable and succulent enough to attract notice. Hernando de Soto, cruising Tampa Bay in 1539, wrote that "beans and pumpkins were in great plenty; both were larger and better than those of Spain; the pumpkins when roasted had nearly the taste of chestnuts." Coronado saw "melons" (probably squash) on a gold-scouting expedition through the American Southwest; Cartier noted "gros melons" (probably pumpkins) in Canada in 1535; and Samuel de Champlain remarked on the "citroules" (squash?) of New England during his voyage of 1605. Columbus's account of his first voyage mentions Cuban fields planted with "calebazzas," or gourds, which were probably hard-shelled squash.

Hernando de Soto, cruising Tampa Bay in 1539, wrote that "beans and pumpkins were in great plenty; both were larger and better than those of Spain; the pumpkins when roasted had nearly the taste of chestnuts."

Recent botanical consensus is that the gourd grown in fifteenth-century Europe (*Lagenaria*) is of Old World origin and a post-Columbian introduction to the Americas. Early references to Peruvian "gourds," creatively used as floats for fishing nets, as well as baskets, buckets, and cooking pots, most likely mean winter squashes. Similarly, early European mentions of "squash"—those, for example, supposedly grown in the Hanging Gardens of Babylon—probably refer to *Lagenaria* gourds. Some of these are edible, and it was likely a tasty *Lagenaria* that the Romans consumed, immature, doused with vinegar and mustard. Roman gardeners also grew gourds for show, in "grotesque forms" up to nine feet long.

Melon—often spelled "million"—was a name commonly applied by newcomers to the American cucurbits, perhaps because the familiar melon was the closest recognizable vegetable equivalent. Melon expectations may have led to some colonial disappointments: Captain John Smith mentions a fruit like a muskmelon grown by the Virginia Indians, only "lesse and worse," almost certainly a squash. The New England settlers were similarly unimpressed by their initial cucurbit contact. The size of the Narragansett *askutasquash* they considered "uncivilized to contemplate" and the squash-and-seafood chowder offered them by hospitable Indian cooks they damned as "the meanest of God's blessings." After the first taste of a Massachusetts winter, however, they rapidly came around, adopting the Indian squash and pumpkins as staple foods. (After all, remarked one Puritan leader astringently, "You must not thinke to goe to heaven on a feather-bed.")

Common practice by the time the *Mayflower* landed was to bake winter squashes and pumpkins whole in the ashes of the fire, then cut them open and serve them moistened with animal fat and maple syrup or honey.

The genial Dutch of New Netherlands found the local *quaasiens* "a delightful fruit," greatly favored by the women because it was easy to cook. Traveller John Josselyn praised the squash in his *New England's Rarities Discovered*, an account of the colonial glories observed in his journeys, published in 1672 at the Sign of the Green Dragon in St. Paul's Churchyard, London. The volume, which includes a flowery poem dedicated to an Indian squaw, describes the squash as "a kind of Melon or rather Gourd for they oftentimes degenerate into Gourds; some of these are green, some yellow, some longish like a gourd, others round like an apple; all of them pleasant food boyled and buttered, and seasoned with spice." Josselyn's *Rarities* also included a Massachusetts "pineapple," which, when picked, erupted into a horde of angry wasps, as well as "Billberries, Black and Sky-coloured," cranberries, tobacco, Indian beans, and the American "Waterlilly," of which the Indians ate the boiled roots. Josselyn tried them and thought they tasted like sheep's liver.

Indian gardens offered a considerable number of varieties of squash. The northeastern tribes grew pumpkins, yellow crooknecks, pattypans, Boston marrows—perhaps the oldest squash in America still in commercial production—and turban squashes; southern tribes raised winter crooknecks, cushaws, and green-and-white striped sweet potato squashes. The Indian name for the fruit, variously rendered as *askutasquash, isquotersquash*, or simply *askoot*, translated as something to be eaten raw, probably the first and least satisfactory means of consumption.

Common practice by the time the *Mayflower* landed was to bake winter squashes and pumpkins whole in the ashes of the fire, then cut them open and serve them moistened with animal fat and maple syrup or honey. The earliest Pilgrim-invented pumpkin pie was a variation on this theme: the top was sliced off a pumpkin; the seeds scraped out; the cavity filled with apples, sugar, spices, and milk; the top popped back on; and the stuffed fruit baked whole. By the next century, the more classic Thanksgiving dinner version, in a crust, had appeared. The "Pompkin Pie" recipe in Amelia Simmons' 1796 cookbook calls for a pudding-like filling of milk, "pompkin," eggs, molasses, allspice, and ginger baked in a "tart paste," or crust, of flour and butter. Yankee culinary ingenuity also devised pumpkin stews and soups (with corn, peas, and beans), sauce (served on meat and fish), porridge, pancakes, bread, butter, and, with much effortful boiling, molasses. Pumpkin beer was brewed from a combination of pumpkin, persimmons, and maple sugar, and pumpkins cut in slices were strung on thread and dried to make pumpkin chips.

The cucurbits also had their nonculinary uses. As early as 1611, a Miss Elizabeth Skinner of Roanoke, Virginia, recommended squash seeds pounded with meal to remove freckles and other unsightly "spottes" from the face. The Indians ate squash and pumpkin seeds as a worm expellant, and whole squash (in quantity) for snakebite. The settlers drank pulverized squash and pumpkin seeds in water for bladder trouble and made tea of ground pumpkin stems to treat "female ills." The various pains of childbirth, toothache, and chilblains were thought to abate if the sufferer chewed on a squash, and the colonists of Jamestown used boiled squash mashed into paste as a poultice for sore eyes.

A hefty number of pumpkins and squashes were needed to supply all these dietary and medicinal needs,

PUMPION-PYE

From Hannah Woolley, The Gentlewomans Companion, *1673*

◆ Take a pound of Pumpion, and slice it; a handful of Thyme, a little Rosemary, sweet Marjoram stripped off the stalks, chop them small; then take Cinnamon, Nutmeg, Pepper, and a few Cloves, all beaten; also ten Eggs, and beat them all together, with as much Sugar as you shall think sufficient. Then fry them like a pancake, and being fried, let them stand till they are cold. Then fill you Pye after this manner: Take Apples sliced thin round ways, and lay a layer of the pancake, and another of the Apples, with Currants between the layers. Be sure you put in a good amount of sweet butter before you close it. When the Pye is baked, take six yolks of Eggs, some White-wine or verjuice, and make a caudle [sauce] thereof, but not too thick; cut up the lid and put it in, and stir them well together, and so serve it up.

and the colonial cucurbit soon outgrew the kitchen garden and was elevated to the status of field crop. It usually sat in its field until October, bulging ripely over the remains of withered vines and stalks, and as such was fair game for the natural disasters recorded in colonial histories as "pumpkin floods." Floods of any kind are rare in October, a notoriously dry month nationwide, but occasional torrential downpours do occur, with accompanying high water and river overflow. Such floods occurred at least twice in the 1780s, once overrunning the pumpkin fields of Maine and New Hampshire, the following year washing out the pumpkins of Pennsylvania and Maryland. Pumpkins, for all their apparent solidity, float, and the unexpected October overflows carried off enough of them that the floods were named for their buoyant orange cargo. In non-flood years, pumpkins were harvested more conventionally, stored in straw in the root cellar, and served up in pies for Thanksgiving dinner, a holiday scornfully referred to by non-participating Episcopalians as St. Pumpkin's Day.

The various pains of childbirth, toothache, and chilblains were thought to abate if the sufferer chewed on a squash, and the colonists of Jamestown used boiled squash mashed into paste as a poultice for sore eyes.

By the 1780s, Yale students, a supercilious crew, were calling New Englanders pumpkin heads, though the term seems to have been inspired as much by their characteristic bowl—or half-pumpkin-shaped haircuts as by their pumpkin-heavy diets. Boston, before it was Beantown, was Pumpkinshire, and New England V.I.P.s, by the mid-nineteenth century, were cozily known as big pumpkins. Unfortunately, by 1845, a pumpkin had also come to mean a stupid and thick-headed person, and has since metamorphosized into bumpkin, an oaf from the boonies.

The pumpkin patch even generated its own certifiable maniac. His name was David Wilbur, born in Westerly, Rhode Island, around 1778. David took to the woods at the age of twenty and spent the rest of his life there, living off the land, in apparent terror of all humankind. He was noted for his uncannily accurate weather predictions (delivered from a safe distance) and his habit of scratching strange signs and figures on field pumpkins, hence his nickname "The Pumpkin Scratcher." He died in 1848. As far as is known, nobody ever deciphered his pumpkin messages.

Europeans were not as taken with the American cucurbits as were their colonial counterparts. The American vegetable introductions were not seriously considered as food until the nineteenth century, except by the long-suffering European livestock, which ate anything. Summer squash reached England in the late seventeenth century, where it was ungratefully dubbed "harrow marrow." (The source of the English *marrow*, meaning squash, is obscure; one guess is that it was thought to have the taste of consistency of bone marrow, a common ingredient in eighteenth-century recipes.) In France, squash seeds stuck in the gullets of the prized Strasbourg geese, destined for pâté de foie gras, and squash in the whole managed to offend the influential horticulturalist Olivier de Serries, who had obtained his garden specimens from Spain. He referred to the new vegetable as "Spain's revenge."

Nonetheless, Americans were undeterred in their enthusiasm for squash. Washington and Jefferson were both squash growers: the Monticello gardens featured pumpkins, "white pumpkins," and "cymlings," the last an early name for the bush scallop or pattypan. (The English called it the custard marrow.) Many new squash varieties were picked up in the nineteenth century by sea captains in the West Indies or South America, and brought back to enrich the gardens of their home ports. By such routes arrived the popular acorn, Valparaiso, Marblehead, pineapple, and Hubbard squashes. The Hubbard squash, cunningly described as "turned up like a Chinese shoe," and said, baked, to taste like a sweet potato, had a long run as America's favorite winter squash. It was formally introduced to American gardens by Marblehead, Massachusetts, seedsman James J. H. Gregory, who traced its homely history in *The Magazine of Horticulture*, December 23, 1857: "Of the origin of the Hubbard squash we have no certain knowledge. The facts relative to its cultivation in Marblehead are simply these. Upwards of twenty years ago, a single specimen was brought into town, the seed from which was planted in the garden of a lady, now deceased; a specimen from this yield was given to Captain Knott Martin, of this town, who raised it for family use for a few years, when it was brought to our notice in the year 1842 or '43. We

Europeans were not as taken with the American cucurbits as were their colonial counterparts. The American vegetable introductions were not seriously considered as food until the nineteenth century, except by the long-suffering European livestock, which ate anything.

were first informed of its good qualities by Mrs. Elizabeth Hubbard, a very worthy lady, through whom we obtained seed from Cpt. Martin. As the squash up to this time had no specific name to designate it from other varieties, my father termed it the 'Hubbard Squash.' " Gregory's business fairly boomed after the acquisition of Hubbard squash seed, and Gregory went on to become something of an authority on squashes, publishing, in 1893, an informative work titled *Squashes: How to Grow Them*.

In the same year, New York seedsman and squash activist Peter Henderson began to inveigh against the pumpkin: "The pumpkin is yet offered in large quantities for sale in our markets, but it ought to be banished from them as it has been for some time from our garden. But the good lieges of our cities are suspicious in all innovations in what is offered them to eat, and it will be many years yet before the masses will understand the modest and sometimes uncouth looking squash is immeasurably superior for all culinary purposes to the mammoth, rotund pumpkin." The masses did indeed continue to swear by their pumpkins, growing such swollen goodies as the orange-yellow Connecticut Field and the Vermont (or Canada, depending on one's side of the border), the darker and flatter Cheese, the warty greenish-black Nantucket, the pear-shaped Tennessee Sweet Potato, and the round blue-gray Possum Nose.

"It will be many years yet before the masses will understand the modest and sometimes uncouth looking squash is immeasurably superior for all culinary purposes to the mammoth, rotund pumpkin."

One advantage of the mammoth pumpkin over the modest squash was its suitability for the carving of jack-o-lanterns. The jack-o-lantern arrived in this country in the mid-nineteenth century along with the influx of potato-starved immigrants from Ireland. An old custom of Ireland and the British Isles, it is said to have originated with a blacksmith named Jack, who sold his soul, at a hefty profit, to the Devil. When the Devil came around to collect, Jack weaseled out of the bargain by sneakily trapping him in a pear tree. This solved matters temporarily, but eventually Jack's irrevocable and final number came up. Barred from the Pearly Gates for all this truck with the Devil, Jack went straight to Hell. The Devil, with the pear tree fresh in his mind, didn't want Jack around either. Just before the gates of Hell shut him out forever, Jack scooped up a burning coal with half of a turnip that he happened, providentially, to be eating. He has used it as a lantern ever since, while wandering around the earth waiting for Judgment Day. In America, the traditional turnip jack-o-lantern rapidly gave way to the enormous and irresistible pumpkin.

Most successful among squashes today is probably the zucchini, which became popular in American gardens in the 1950s, re-introduced to this continent from Italy. The Italians acquired the zucchini in mysteriously undocumented fashion over three hundred years ago, and formed an immediate affinity for it. The name is a derivation of an Italian word meaning sweetest. Six summer squash plants, one expert figures, yield an average of fifty pounds of fruit per summer, a conservative estimate for the legendarily prolific zucchini. Home garden production, in fact, occasionally reaches such heights that in some parts of the country people are advised to lock their cars in midsummer to prevent vegetable desperadoes from dumping loads of zucchini in the back seat. To cope with the explosive zucchini patch, often a vegetable version of "The Sorcerer's Apprentice," entire cookbooks have been devoted entirely to zucchini. Recipes include zucchini quiches, omelets, pickles, cakes, cookies, and curries, zucchini marmalade, zucchini stroganoff, and even zucchini-and-peanut-butter sandwiches.

An old custom of Ireland and the British Isles, the jack-o-lantern is said to have originated with a blacksmith named Jack, who sold his soul, at a hefty profit, to the Devil.

Much of the squash breeding effort in this country over the past twenty years has been devoted to zucchini— which is now available in a rainbow of colors other than the standard flecked green. The zucchini appears internationally on the tables of France, as the *courgette*; England, as the *baby marrow*; and Spain, as the *calabacin*. Still, it's not a squash for everybody. "The first zucchini I ever saw," says author John Gould, "I killed it with a hoe." Mr. Gould is also down on the pattypan ("a cross between a Scottish curling stone and the end cut from a roast of foam rubber") and feels, in fact, that the best use for all squashes is to dry them and hang them from the trees as bird houses.

A perhaps more acceptable, since less squashlike, alternative is the spaghetti squash, or vegetable spaghetti, a variety of *C. pepo* whose insides, when baked or boiled, unravel into a mass of fine, spaghetti-like strands touted as a low-calorie substitute for pasta. The spaghetti squash—a hard-shelled winter squash shaped roughly like a football—was originally cultivated in Italy and Spain. At the beginning of this century, samples of the Italian version were obtained by the seed company of T. Sakata in Japan, which produced the smaller, but more productive, hybrid strain predominant today.

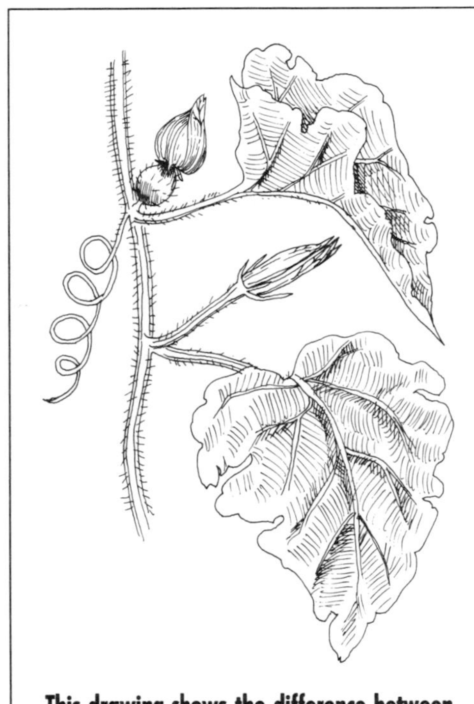

This drawing shows the difference between the female bud, with its round ovary, and the male bud, with its long, thin stem.

Adventurous squash eaters do not necessarily confine themselves to the conventional fleshy fruits. Squash flowers, especially those of zucchini, pattypan, and summer crookneck squashes, are both edible and flavorful. Friar Bernardino de Sahagún reported squash blossom hors d'oeuvres at Montezuma's banquet table, though he was regrettably vague about their manner of preparation. These days the blossoms are sautéed, dipped in batter for fritters, or stuffed with rice and meat. The Zunis of the Southwest traditionally ate squash blossoms in soup, choosing the large male flowers, which were considered the most delectable.

Prospective soup cooks have a choice between male and female flowers because monoecious plants, such as squashes, pumpkins, cucumbers, and melons, bear both. Dioecious plants, in contrast, segregated like Victorian boarding schools, bear either male or female flowers, never both. This seeming botanical propriety is nature's way of preventing self-fertilization and promoting the beneficially varied genetic scrambling that is the point of sexual reproduction in the first place. Spinach, asparagus, and holly are all dioecious, which means that to get red berries for your Christmas wreaths, you need a breeding pair of trees. In monoecious plants, often the male and female flowers mature at different times to encourage cross-fertilization. In squashes, for example, the male flowers, the pollinizers, open first. Male and female flowers are simple to tell apart: the males have straight skinny stems leading directly to the bud; females have a prominent bulge at the top of the stem adjacent to the petals, containing the ovary. Upon fertilization, this ovary develops into a mature squash or pumpkin—gaining, in the more spectacular cases, as much as eight pounds a day.

Squash and pumpkin blossoms are borne on indefatigable vines that, given their head, will happily overrun one hundred square feet or more of garden. This insidious habit, of particular concern to gardeners with limited growing space, has long been a target of breeders and plant scientists. The results of their professional manipulations are known as bush cultivars, plants in which the internodes—the lengths of stem between leaves—have been drastically shortened. The truncated cultivars take up a quarter, or less, the space of the standard vines, but in many cases have been found to bear smaller and fewer fruits than their unconfined relatives. A possible reason for this, researchers suggest, is the reduced photosynthetic area that follows reduction in vine length: shorter vines mean fewer leaves, which in turn means less sunlight-derived energy to fuel the development of fruit. To owners of pocket-handkerchief-sized vegetable plots, however, a small pumpkin is better than no pumpkin at all. Still, even a bush pumpkin can swallow sixty square feet of garden space, so genetics has yet to give us a jack-o-lantern in a flowerpot.

The pumpkin has been immortalized in prose and poetry by such literary greats as Mother Goose, Washington Irving, John Greenleaf Whittier, and L. Frank Baum, and few autumns pass without somebody quoting James Whitcomb Riley's colloquial tribute to the time "when the frost is on the punkin." Henry David Thoreau tossed it a left-handed compliment, reflecting on the solitary banks of Walden Pond that he would rather sit on a pumpkin (alone) than on a velvet cushion (crowded), and in 1697, master-storyteller Charles Perrault provided his Cinderella with the best of both worlds, transforming a pumpkin into a velvet-cushioned coach to carry her in cucurbital elegance to the prince's ball. Best of the pumpkin tales, however, is one of Aesop's fables, which tells of a man who lay beneath an oak tree, criticizing the Creator for hanging a tiny acorn on a huge tree, but an enormous pumpkin on a slender vine. Then, the story goes, an acorn fell and hit him on the nose.

Squash and pumpkin blossoms are borne on indefatigable vines that, given their head, will happily overrun one hundred square feet or more of garden.

CUCUMBERS

The garden cucumber, *Cucumis sativus*, comes to us by way of India, where it has been cultivated for at least three thousand years—and perhaps considerably longer, since excavations in 1970 at Spirit Cave on the Burma-Thailand border dredged up seeds of cucumbers, peas, beans, and water chestnuts, remains of meals eaten, according to radiocarbon dating, in 9750 B.C. The wild ancestor of our present-day edible cucumber has never been identified. The best guess seems to be *C. hardwickii*, an unappetizing native of the Himalayas, small and bitter, scattered with nasty little spines. It may have been *C. hardwickii* that the unfortunate Enkidu ate along with worms, figs, and caper buds in the ancient Sumerian epic *Gilgamesh*. Time and human effort, however, eventually created a sweeter and less off-putting vegetable, and the result quickly spread. The Egyptians ate them at every meal, dipped in bowls of brine, and used them to make a questionable drink called cucumber water. To do this, a hole was cut in a ripe cucumber, the inside stirred up with a small stick, the hole plugged, and the cucumber then buried in the ground for several days. When unearthed, boasts an ancient recipe, "the pulp will be found converted into an agreeable liquid," possibly the concoction the Israelites mourned as they slogged thirstily through the desert after Moses.

Even so, cucumber opinion seems to have been somewhat divided: a quote from the *Apocrypha* states, "A scarecrow in a garden of cucumbers keeps nothing," which certainly implies that the cucumber was not the top crop on the block. Nutritionally, the "nothing" is literal. The average cucumber is 96 percent water, with not much else other than a paltry seventy International Units of vitamin A (1½ percent) of the Recommended Daily Allowance), all in the peel. At that rate, in terms of vitamin A, it takes 120 unpeeled cucumbers to equal 1 carrot. Food historian Waverley Root describes the cucumber as "as close to neutrality as a vegetable can get without ceasing to exist."

Still, all that water gave the characteristically cool cucumber a banner reputation as a refreshing thirst-quencher. Early caravans took them along as a sort of vegetable water bottle; overheated Greeks mashed them and mixed the pulp with honey and snow to make an ancient version of sherbet. The Romans were enthusiastic about them, occasionally eaten raw, but more often boiled and served with oil, vinegar, and honey. The Emperor Tiberius was positively mad for them, consuming, according to Pliny, ten a day, every day. To indulge such autocratic whims, Roman gardeners developed methods of growing cucumbers in earth-filled baskets, carting the plants from spot to spot to make the most of limited off-season sun. By the first century A.D., cucumber frames had been devised, of the sort Peter Rabbit fell into so disastrously in Mr. McGregor's garden, glazed with translucent sheets of *specularia*, which was probably mica.

The Emperor Tiberius was positively mad for cucumbers, consuming, according to Pliny, ten a day, every day.

Cucumber cultivation dwindled with the disintegration of the Roman Empire, and only reappeared in force in the sixteenth century, a period that Bert Greene refers to as the "cucumber black-out." During this dark interim, there were a few patches of horticultural enlightenment. In the late eighth century, Pepin (the Wise) of France, possibly influenced by a classical belief that steeping seeds in cucumber juice protected them from insect predation, ordered three rows of cucumbers planted around his vineyards to ward bolls, borers, and cutworms off the valuable grapes. Pepin's renowned son, Charlemagne, following praiseworthy parental example, ordered cucumbers planted in the royal garden, and even declared the cucumber his favorite fruit. He reportedly ate them for dessert, in custard tarts.

The cucumber made its appearance in England, the story goes, during the reign of Henry VIII, under the urging of Catherine of Aragon, who liked them, sliced, in her Spanish salads. It lasted better than Henry's wives, and by the reign of Elizabeth I, English gardens boasted five varieties: the Common, the Turkey, the Adders, the Pear Fashion, and the Spanish. John Parkinson mentioned seven varieties in 1629, stating on the side that, "in many countries they use to eat coccumbres as wee doe apples or Peares," but cucumber suspicion persisted in some quarters. On September 22, 1663, Samuel Pepys recorded in his diary, "this day Sir W. Batten tells me that mr. Newhouse is dead of eating cowcumbers, of which the other day I heard of another, I think." Medical authorities confirmed that cucumbers filled the body with "cold noughtie humors" and brought on ague. The external cucumber produced more positive effects: sleeping on a bed of them was said to cure fever— hence the saying "cool as a cucumber"; cucumber leaves stamped in wine were used to treat dog bites; women wishing for children were encouraged to wear a cucumber suggestively suspended from the waist; and to dream of cucumbers was believed to indicate the imminence of falling in love. The Romans prosaically claimed that cucumbers scared away mice, and John Gerard claimed that, eaten three times a day in oatmeal pottage, they could cure swellings of the face, noses "red as red Roses," pimples, "pumples," and like disasters of the seventeenth-century complexion.

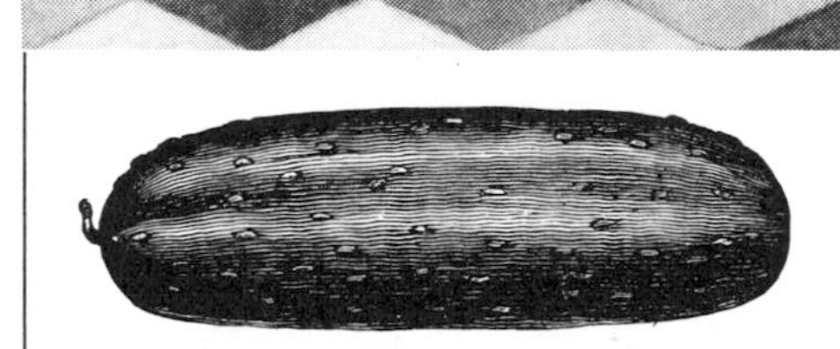

On September 22, 1663, Samuel Pepys recorded in his diary, "This day Sir W. Batten tells me that mr. Newhouse is dead of eating cowcumbers, of which the other day I heard of another, I think."

The horticultural cucumber was subject to an equal array of arcane beliefs. Cucumbers were said to be frightened of thunderstorms, so one expert gardener advised draping the plants in comforting "thin Coverlets" in the event of violently inclement weather. In the sixteenth century, garden advisors Etienne and Liebault announced that cucumber seed over three years old would yield radishes when planted (they also suggested crushing flat parsley with a garden roller to make it curly), and a number of growers who should have known better claimed that cucumbers waxed and waned along with the moon. It was customary to pick cucumbers at the full of the moon, in hopes of getting the very biggest, which were also considered the very best. Batty Langley objected to this practice in his 1728 *New Principles of Gardening*: " 'Tis a very great Custom amongst a great many People to make choice of the very largest Cucumbers, believing them to be the best, which are not, but instead thereof, are the very worst, except such as are quite yellow. Therefore in the Choice of Cucumbers, I recommend those that are about three Parts grown, or hardly so much, before those very large ones, whose Seed are generally large, and not fit to be eaten, excepting by such Persons whose stomachs are very hot. . . ." And the vitriolic Dr. Samuel Johnson objected to all cucumbers, of whatever size, saying, "A cucumber should be well sliced, and dressed with pepper and vinegar, and then thrown out, as good for nothing." It's possible that he may have changed his mind given a chance at cucumbers prepared à la Elizabeth Rafald, who, in *The Experienced English Housekeeper* (1769), recommended that large-sized cucumbers be stuffed with partly cooked pigeons, cleaned, but with heads and feathers left on, so that the heads appeared attached to the cucumber. These were cooked in broth and served garnished with barberries.

Heinz started bottling pickles in the 1870s, and by 1888 had a twenty-two-acre factory complex outside of Pittsburgh, complete with steam heat, electric lights, "equine palaces" for the 110 jet-black Heinz horses (who pulled cream-colored wagons trimmed in pickle-green), and a 1,200-seat auditorium with a stained glass dome.

Cucumbers came to the western hemisphere with Columbus, who planted them in his experimental garden of 1493. They seem to have done well. By 1535, Jacques Cartier observed "very great cucumbers" in Canada, and de Soto found "cucumbers better than those of Spain" in Florida in 1539. The colonists planted them—three shillings-worth of cucumber seeds appeared on young John Winthrop's seed bill—and by 1806 Bernard M'Mahon listed eight standard varieties in American gardens, including the Long Green Turkey, twenty inches long at maturity.

Also grown, according to M'Mahon, was the West Indian gherkin, *Cucumis anguria*, nicknamed the Jerusalem pickle. *C. anguria* was described in an eighteenth-century natural history of Jamaica as a walnut-sized pale green fruit, "far inferior" to the garden cucumber, but still edible if soaked in vinegar. It was used accordingly, since pickles were a prime use of the colonial cucumber. In the absence of alternative preservation techniques, the colonists pickled practically everything, from walnuts and peaches to eggs and artichokes, using processes not noticeably different from those used today. Amelia Simmons pickled her cucumbers in white wine vinegar, with added cloves, mace, nutmeg, white peppercorns, "long pepper," and ginger; Harriott Pinckney Horry's *Receipt Book* lists a similar recipe,

"To Mango Muskmellons and Cucumbers and to pickle French Beans, girkins, etc.," also said good, with a little adaptation, for oranges.

Preservation by pickling works by immersing the food in an acid solution—most commonly vinegar—which prevents the growth of microorganisms and accompanying food spoilage. Vinegar is an invention of unspecified but considerable antiquity. The Babylonians and the ancient Chinese had it, and the Spartans used it in their notorious black broth, a mix of pork stock, vinegar, and salt that, according to the Athenians, was enough to make anyone willing to die. The Roman legions routinely used vinegar to purify their drinking water. Our word *vinegar* is derived from the French *vin aigre*, or "sour wine," reflecting its first major source, as a by-product of the winemaking industry. Wine goes sour under the ministrations of a bacterium called *Acetobacter*, which consumes the existing alcohol, leaving behind a mixture of 4 percent acetic acid in water. A similar bacterial process sours beer to yield malt vinegar, and apple juice to yield the American specialty, cider vinegar. In the absence of any kind of vinegar, frontier families pickled their produce in the ever-available whiskey.

The most famous pickle in American history was that of Henry J. Heinz, which, accompanied by a blue-eyed tot clutching a can of baked beans, dominated the national scene for decades. Heinz started bottling pickles in the 1870s, and by 1888 had a twenty-two-acre factory complex outside of Pittsburgh, complete with steam heat, electric lights, "equine palaces" for the 110 jet-black Heinz horses (who pulled cream-colored wagons trimmed in pickle-green), and a 1,200-seat auditorium with a stained glass dome. Sightseers got a guided tour and a souvenir pickle in plaster. Pickles were endlessly popular in the nineteenth century, often the only taste relief in a monotonous diet of meat and potatoes. "A dinner or lunch without pickles of some kind," stated *Good Housekeeping* magazine in 1884, "is incomplete." On the better tables, the essential pickles may even have been served up with a runcible spoon, a term coined in 1871 by nonsense poet Edward

Lear for what was actually a pickle fork with three broad curved prongs. (Lear's immortal Owl and Pussycat used theirs to eat mince and slices of quince.)

American processed-pickle consumption has increased greatly since Heinz's death in 1919—up nearly eightfold nowadays, to over eight pounds of pickles per person per year. Heinz, who, along with his cucumber pickles put out pickled onions, pickled cauliflower, and cider vinegar produced in his own vinegar plant, drew up his first cucumber contract with the pickle growers of La Porte, Indiana, in 1880. In it, he agreed first to supply the seed, then to buy the harvest. Improvements in the bottled product, Heinz held, must start in the ground.

Until the Heinz era, little effort had gone into the improvement of the cucumber, in the ground or otherwise. The first notable deliberate attempt at cucumber hybridization resulted in Tailby's Hybrid, a high-yield large-fruited cucumber introduced to gardens in 1872. It is carried in Burpee's 1888 catalog, along with nineteen other cucumber varieties, including the Russian or Khiva Netted cucumber, oval with a white-netted brown skin, said to be "well adapted for cold, bleak situations," and the Serpent or Snake cucumber, which grew up to six feet in length, coiled like a snake. Vilmorin-Andrieux describe it in the 1896 edition of *The Vegetable Garden*, adding that when ripe it exudes "a strong odour of Melons." Among the other twenty-seven listed cucumber cultivars is the Bonneuil Large White, a sweetly scented, ovoid, and oddly angular cucumber grown near Paris exclusively for use in perfumes.

Vilmorin-Andrieux suggest that growers straighten their market cucumbers ("as one good and straight Cucumber is worth nearly a dozen small and deformed ones") and describes a method for doing so, by forcing the young fruits into open-ended cylindrical glasses approximately 12–15 inches long and 1½–2 inches in diameter. Over a thousand workers were employed in this task in one English market garden, say Vilmorin-Andrieux, which somehow brings to mind such horticultural bizarrities as Lewis Carroll's gardeners painting the roses red in Wonderland. Those cucumbers too far gone in deformity were brutally sent to the pickle factory. The obsession with straightening the cucumber is an ancient one—the early Chinese growers suspended stones from the ends of fruits with a tendency to curl—and modern breeders have selected for straightness, along with such traits as size, yield, disease resistance, flavor, and the commercial bugbear, shelf life. Innumerable cultivars are available today, including many time-honored heirloom breeds, among them the Lemon, which looks like one; the White Wonder, a dull ivory color; the tiny Crystal Apple; and the blimp-shaped Zeppelin.

Flavor in garden cucumbers has been continually plagued by bitterness, the occasional mouthful of which still gives a chilling reminder of what cucumber-eating was like in the bad old days of the prehistoric Himalayan wilds. The bitterness is due to a class of compounds called cucurbitacins, terpene derivatives, that are as repulsive to certain insect pests as they are to human beings. Cucurbitacin-less mutants have been developed, enabling growers to produce crops of non-bitter cucumbers, but the tastier fruits, chemically disarmed, are consequently more susceptible to insect damage. Kitchen lore holds, unreliably, that cucumbers can be de-bittered by slicing off an end and then rubbing the cut surfaces together. More effective is the method described by Bert Greene, who recommends peeling the potential offender and placing it in an earthenware bowl with a bit of salt and sugar, plus a dollop of vinegar, for half an hour.

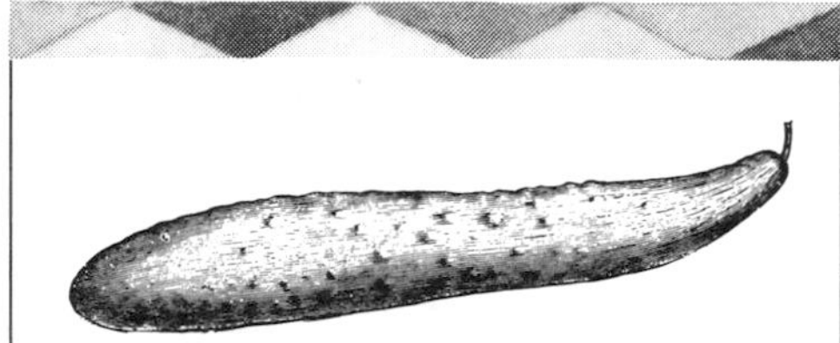

The obsession with straightening the cucumber is an ancient one—the early Chinese growers suspended stones from the ends of fruits with a tendency to curl—and modern breeders have selected for straightness.

Cucumbers, like melons, squashes, and pumpkins, are monoecious. The flowers, showy yellow-orange numbers with five-lobed corollas, either possess five (male) stamens or a single (female) pistil with three stigmas. The males usually grow in clusters; the females are loners. Many pickling cultivars these days, however, are sexual oddities. Botanically known as gynoecious plants, these vines bear only female flowers, and consequently produce unusually heavy crops. Even odder are the parthenocarpic cucumbers, which seem to have escaped the sexual rat race altogether. These set fruit without benefit of pollination, and thus produce no seeds. To succeed at this, the plants must be closely protected from invading pollen, a difficult enough feat to quadruple the price of the seedless offspring. Hybrids between gynoecious and parthenocarpic plants can be grown in the ordinary garden, still produce seedless fruits, and are perhaps the best bets for the seedless future.

As well as de-sexing the cucumber, breeders have made them small. The standard cultivar nabs for itself four square feet or so of garden living space; bush cultivars can occupy less than half that, though space-saving gardeners pay the price in reduced yields. The average standard cultivar, one study shows, produces 4.43 pounds of cucumbers per square foot of plant; the average bush model produces 3.32 pounds per square foot. Averages being what they are, this can mean as few as ten to as many as sixty cucumbers per vine.

The caretaker, according to the unappreciative Gulliver, "had been eight years upon a project for extracting sun-beams out of cucumbers, which were to be put into vials hermetically sealed, and let out to warm the air in raw inclement summers."

Gulliver, on his post-Lilliput voyage to Laputa, ran into what may have been the first of the scientific cucumbers. Their caretaker, according to the unappreciative traveller, "had been eight years upon a project for extracting sun-beams out of cucumbers, which were to be put into vials hermetically sealed, and let out to warm the air in raw inclement summers." It sounds less far-fetched nowadays, though to date no one seems to have put a hand to it. Cucumbers have, however, been grown in space, by the Soviet cosmonauts who broke the world space endurance record in 1982 by spending 211 days on board the space station Salut-7. The cooped-up crew spent some of their time cultivating cucumbers, radishes, and lettuces in the spaceship kitchen garden.

On the opposite end of the scale, cucumbers have been grown nearly a mile beneath the earth's surface, at the Inco Ltd. Creighton nickel mine at Copper Cliff, Ontario, Canada. The project, cooperatively supported by the mining company, Laurentian University, and the Canadian government, is designed to ascertain the potential of underground food production, particularly for northern communities that are ordinarily unable to obtain fresh produce during the winter months. The mine, naturally heated from the subterranean depths, is about 84 degrees F at the 4,600-foot "garden" level—a three-minute trip by elevator—where pine seedlings and cucumbers flourish in individual black plastic bags stuffed with peat moss and vermiculite. Researchers have also successfully grown underground lettuce and tomatoes.

While cucumber-growing for those who have a nickel mine looks to be a year-round proposition, for most of us cucumber season consists of a few pitifully short weeks in the summer. For those with a passion for cucumber, there are season-stretching edible alternatives, in the form of cucumber-flavored herbs. The leaves of borage (*Borago officinalis*) and salad burnet (*Poterium sanguisorba*) both taste vaguely of cucumber and are highly recommended for salad lovers. Jefferson grew salad burnet, a ferny blue-green perennial, in quantity at Monticello. In his day, sprigs of burnet were used for flavoring in glasses of wine.

The last in the list of edible cucumbers is not a cucumber at all. The sea cucumber, which is shaped like one, is an echinoderm, relative of the starfish, the sea urchin, and the less familiar feather-star and brittle-star. All are radially, rather than bilaterally, symmetrical, and all are pentamerous, though the sea cucumber's five-part construction is largely internal. The outer animal looks like a fat worm and spends its life shuttling unexcitingly back and forth across the sea bottom, swallowing sand and extracting from it anything digestible in the way of organic matter. Some sea cucumbers are gaudy—purple, orange, and yellow, as well as a lurid cucumber-green—and all are tough, many surviving undersea pressures that would flatten a lesser organism. Over six miles down, on the bottom of the mid-Pacific Philippine Trench, almost the only inhabitants are the indomitable sea cucumbers. On an international scale, they're less popular than *C. sativus*, but some brave souls do eat them.

MELONS

Our word *vegetable* comes from the Latin *vegere*, meaning, in spite of the sedentary implications of vegetate, to animate or enliven; *fruit*, on the other hand, derives from *frui*, meaning to enjoy. Most people do enjoy fruit, in the culinary (not botanic) sense of snack food or dessert, which is why teachers get apples instead of eggplants, and why Hades chose a pomegranate to tempt Persephone. Fruits generally have the edge over vegetables because they appeal to the voracious human sweet tooth—a sense actually located at the tip of the tongue and active from infancy. Babies, no fools, repeatedly opt for applesauce over strained peas, and the sweetness preference persists undeterred into adulthood. Most temperate-zone fruits contain about 10–15 percent sugar by weight; tropical-zone fruits, at 20–26 percent sugar, are even sweeter. In the annual vegetable garden, the melons are a bit on the low side: muskmelons, when ripe, contains 6–8 percent sugar, which means that the average melon contains only half as much sugar as the average apple or pear. Watermelons—described by Mark Twain as "chief of this world's luxuries . . . when one has tasted it, he knows what angels eat"—total 6–12 percent sugar.

Melons, all basically classed as *Cucumis melo*, are organized by taxonomists into three major groups, which unfortunately correspond only vaguely to common garden melonspeak. The melon known to most of us as the cantaloupe is botanically a muskmelon, *C. melo* var. *reticulatus*, the *reticulatus* referring to the netted shell or rind. The fashion-conscious French call these embroidered melons. They are heavily fragrant melons, green- or orange-fleshed, usually weighing two to four pounds apiece. The true cantaloupe, *C. melo,* var. *cantalupensis*, is commonly segmented, has a hard, but not netted, shell and usually orange flesh. These are grown primarily in Europe. The name comes from

Cantalupo ("wolf howl") in Italy, site of a palatial papal vacation home outside Rome, where the melons were reputedly first cultivated in Europe in the sixteenth century. Casaba and honeydew melons are winter melons, scientifically known as *C. melo* var. *inodorus* because they lack the intense fragrance of other melons. All three of the melon groups—muskmelons, cantaloupes, and winter melons—are closely related and, given a sporting chance, will promiscuously interbreed. Odd man out is the watermelon, *Citrullus vulgaris*, prevented by genetic distance from crosspollinating with other garden melons.

The watermelon is a native of tropical Africa; Dr. Livingstone, before his retrieval by Stanley, observed them growing wild in the course of his jungle jaunt on the Dark Continent. The *Cucumis* melons, on the other hand, came originally from Persia. The ancient Greeks ate them. One story recounted by Waverley Root tells of a melon-half pitched by a heckler at Demosthenes in the course of a political debate. Demosthenes, never at a loss, is said to have promptly clapped the melon on his head and thanked the thrower for finding him a helmet to wear while fighting Philip of Macedonia. In the first century A.D., Pliny, who seems never to have had one pitched at him, describes the classical *melopepo*, a round, yellow fruit that spontaneously detached from its stem when ripe, probably the Roman version of the garden melon.

The melon was finally cultivated again in quantity in fifteenth-century France, where it enjoyed a popularity craze of the sort that these days surrounds rock musicians.

The melon faded from Europe with the decline of the Roman Empire. It reappeared sporadically, following contacts with the Near East—Charlemagne is said to have acquired some in the course of his altercations with the Spanish Moors, and the Crusaders may have brought a few home from their debilitating struggles with the Saracens. The melon was finally cultivated again in quantity in fifteenth-century France, where it enjoyed a popularity craze of the sort that these days surrounds rock musicians. By 1583, the Dean of the College of Doctors of Lyons, Professor Jacques Pons, had solemnly produced a *Succinct Treaties on Melons*, which listed fifty different methods of melon preparation; and in 1699 John Evelyn, in superlatives usually reserved for salad greens, deemed melons "the noblest production of the garden." Even Montaigne, who generally preferred salted beef to sweets, fell for the melon, announcing in the course of his celebrated *Essays*, "I am not excessively fond of salads nor of fruits, except melons." Out-of-season melons were eagerly cultivated in hotbeds and greenhouses, and Louis XIV, never one to stint himself, had seven varieties cultivated at Versailles, all under glass.

Despite a number of spurious sightings, there were no melons in North America until the European colonists brought them here. Early records of Indian "pompons" probably refer to less luscious pumpkins or squash. One of the earliest accounts of the colonial melon came from Adrien Van der Donck, a Dutch administrator who arrived in New Netherlands in 1642 with the doubtful assignment of keeping an eye out for the interests of the financial bigwigs back home. Van der Donck, who had the distinction of being both the only lawyer and the only university graduate in the Dutch colonies, was an avid observer and notetaker, and in 1653 published the results of both activities in his comprehensive *Description of the New Netherlands*. In it, he describes the colonial muskmelons, which, nicknamed Spanish pork, grew "luxuriantly."

The Dutch also had watermelons, called citrulls or water-citrons, which Van der Donck refers to as "a fruit only known before to us from its being brought occasionally from Portugal." The colonial versions were sometimes crushed for juice, a popular beverage—and the singleminded English fermented it, to make watermelon wine. John Josselyn, in *New England's Rarities*, mentioned its use as an antidote for fever: "watermelon is often given to those sick of Feavers and other hot diseases with good success." (Failing that, home medics advised cranberry conserve or sassafras chips boiled in beer.) Josselyn thought the watermelon "proper to the countrie," which it wasn't.

The colonial watermelon was at the small end of the scale by present standards; nowadays the large number of available cultivars range in weight from a tiny five to a Sunday-school-picnic-sized hundred pounds.

WATERMELON CAKE

From Smiley's Cook Book and Universal Household Guide, *1896*

½ cup butter
1 cup sugar
½ cup sweet milk
3 whites of eggs
2 cups flour
1 teaspoon cream tartar
½ teaspoon soda
Flavor with lemon

Take a little more than ⅓ of the mixture and to it add 1 teaspoon liquid cochineal and ½ cup raisins. Put the red part in the center and bake. Cover with a frosting colored green with spinach.

Also, despite the lurid pink associated with the word *watermelon*, the generic modern watermelon is red-fleshed and oval-shaped with a green-on-green striped rind. Contrary to supermarket expectations, this summer special is far from the only watermelon—*C. vulgaris* can be round, ovoid, oblong, or practically cylindrical, with rinds of pale green to almost black, often striped or marbled, and flesh colored orange, yellow, or white as well as pink and red. Late nineteenth-century catalogs routinely offered twenty to thirty watermelon cultivars, among them the mouthwatering Golden Honey (yellow-fleshed), the Jersey Blue (dark blue rind), and the Moon and Stars (deep green rind sprinkled with heavenly bright yellow spots). Low-sugar varieties—citrons—were grown specifically for pickling, and, judging by Amelia Simmons, watermelon "rine" pickles were common at least by 1796.

Despite the lurid pink associated with the word *watermelon*, the generic modern watermelon is red-fleshed and oval-shaped with a green-on-green striped rind.

Pickling melons are rare in the sugar-laden present, but the twentieth century does boast seedless melons, a Japanese development that bids fair to eliminate old-fashioned summer seed-spitting orgies on the back porch. Seedlessness is a negative in some circumstances: the Chinese for centuries have enjoyed watermelon seeds preserved in salt, and in the Near East, watermelon seeds, roasted, are sold in bags like popcorn. They may be the snack of choice for Type A behavers. The *American Journal of Medical Science*, circa 1926, claimed that watermelon seeds contained a substance effective in the control of hypertension. The seedless watermelon, first introduced to this country in 1948, is triploid, which means that instead of the ordinary two sets of chromosomes per cell, these variants have three. The triploidy also explains all the *tris*, and *triples* that pop up in seedless watermelon varietal names. The flesh, extra-solid and sweet, is generally thought superior to that of the seeded melons, but seedless melon seeds, unsurprisingly, are expensive, costing five to six times more than those of their diploid relatives; yields are lower, and they must be planted in proximity to ordinary melons for crosspollination.

Like the early watermelon, the early muskmelons and associates were, to the critical modern eye, hopelessly tiny. The Romans, who imported their melons from Armenia, were dealing in fruits the size of oranges, and even the green-fleshed Nutmeg, so popular in the early nineteenth century, started out not much bigger than a softball. None of the early melons was as sweet as their twentieth-century descendants, though the beleaguered South during the Civil War boiled melons down as a source of sugar and molasses. By the Civil War years, green-fleshed melons gradually were being replaced in the popular favor by orange-fleshed melons, which were both gaudier and higher in vitamin A. The interest in novelty melons picked up toward the end of the century. One such, the Banana melon or Banana cantaloupe, was introduced in 1883, a smoothly elongated pale yellow fruit with salmon-pink flesh. "When ripe," commented seedsman J. J. H. Gregory, of Hubbard squash fame, "it reminds one of a large overgrown banana and, what is a singular coincidence, it smells like one." It doesn't seem to have tasted particularly good, and W. Atlee Burpee, who states significantly that they grow it in New Jersey, damns it as "poor quality." Burpee, who seems critical of foreign novelty melons, nonetheless offers a selection of them in very small and undescriptive print, repressively reminding the buyer that "our American sorts are best adapted for market purposes."

The Romans, who imported their melons from Armenia, were dealing in fruits the size of oranges, and even the green-fleshed Nutmeg, so popular in the early nineteenth century, started out not much bigger than a softball.

Professional melon-breeding similarly took off at the end of the nineteenth century, with some fifty new melon varieties appearing on the market between 1880 and 1900. The casaba melon—named for Kasaba, Turkey—arrived in this country in the 1850s under the auspices of the U.S. Patent Office, which distributed packets of free seeds to potential growers. However, no one seems to have liked it much—the casaba was green, in a prevailing atmosphere of orange—and it only became popular in the 1920s. The honeydew melon, also green, appeared in the early 1900s, discovered, the story goes, by a gentlemen named Gauger who squirreled away the seeds of the melon served him for breakfast at his New York City hotel. The melon was subsequently identified as a French winter melon called White Antibes.

Most recent melon-breeding has concentrated on the development of disease-resistant varieties, though modern science has done little about one of the foremost melon pests, the human being. Even Gilbert White, the mild-mannered English clergyman who authored the famous *Natural History of Selborne*, had his share of such melon troubles. His gardening diary of the 1760s is a chronicle of familiar woes, beginning with his French beans, "strangely devoured" by snails; his grapes, attacked by wasps (White diverted them by strategically hanging up sixteen bottles of treacle and beer); his cabbages, gnawed by hares; and his turnips and celery, withered by drought. The night of September 18, 1764, was the final straw: his cucumbers and melons—and his horse block—were pulled to pieces by persons unknown. To Reverend White's credit, he never resorted to drastic measures, of which there were many available. In *The Encyclopaedia of Gardening* (1822), John Loudon lists under "Machines for destroying Vermin and for Defense against the Enemies of the Garden" two "engine traps for man," the common and the humane, both of which in more or less merciful fashion broke the legs of unauthorized intruders. Luckily for American literature, such were not employed by the farmers of Hannibal, Missouri, back in the days when Samuel Clemens—not yet Mark Twain—was filching watermelons. The first watermelon that Sam Clemens stole—"retired" is the word he preferred, as in "retired from circulation"—he took to a nearby lumberyard, broke open with a rock, and found green.

The honeydew melon appeared in the early 1900s, discovered, the story goes, by a gentleman named Gauger who squirreled away the seeds of the melon served him for breakfast at his New York City hotel.

What should a high-minded young man do after retiring a green watermelon? What would George Washington do? Now was the time for all the lessons inculcated at Sunday School to act.

And they did act. The word that came to me was "restitution."

So young Clemens took his green watermelon back to the farmer and conned him into apologizing and handing over a ripe one.

This morally reprehensible course is about the best available to the possessors of green melons, since a melon, once picked, is as sweet as it is ever going to get. The sugar content of melons, like that of oranges and pineapples, depends on the length of time spent attached to the stem and the photosynthesizing leaves.

Melons do not accumulate starch, which means that there is no comfortable backlog of sugars in storage available for release after picking. The melon itself is programmed to self-pick once sugar content reaches optimal levels. At this point, a separation, or abscission, layer forms across the stem, blocking the further downward passage of nutrients, and effectively booting the mature melon out of the parental nest. The mature melon literally falls off the vine into the hand of the properly patient gardener. If wrenched off prematurely, as is often the case with the commercial muskmelon, which ships poorly when fully ripe, there's not much that man or nature can do. Ethylene gas has little effect on the stubborn muskmelon; experiments have shown that it can induce softening of the unripe flesh, but no increase in sugar or flavor components.

Not all melons appeal to growers on the grounds of sugar content. *C. melo* var. *chito*, the mango melon or vine peach, has a hard yellow shell and pale unscented flesh, and was grown in the nineteenth century as a pickling melon. A recipe of the 1890s recommends it as a dinner dish, stuffed with cabbage and baked. Less edible are *C. melo* var. *flexuosus*, the sinuous snake melon, and *C. melo* var. *dudaim*, nicknamed Queen Anne's pocket melon or the pomegranate melon. Both are purely ornamental; the ball-sized, yellow-and-orange pomegranate melon, a relative of the luscious cantaloupe, is so fragrant that it was once carried by upper-class ladies as a pomander ball.

Melon seeds, if swallowed, pass peacefully through the human digestive tract and do not, no matter what insensitive friends tell you, sprout and grow in the human stomach. Ladies who look like they've swallowed watermelons haven't.

The first watermelon that Sam Clemens stole—"retired" is the word he preferred, as in "retired from circulation"—he took to a nearby lumberyard, broke open with a rock, and found green.

CORN

The number-one crop in traditionally corny Kansas, home of Dorothy, Toto, and Dwight David Eisenhower, is wheat. Number two is sorghum, and corn limps in there in third place, just ahead of soybeans. Still, Kansas is part of the Corn Belt, the band of corn-rich cropland that has been steadily oozing westward since colonial days. The original American Corn Belt extended down the Atlantic Coastal Plain, from Massachusetts to Georgia. It then shifted into the Piedmont, and later hopped the Appalachians and spread across Kentucky and Tennessee. Tennessee raised most of the nation's corn well into the nineteenth century, when the Belt moved again, to the territory it occupies today: fifty million acres stretching from Ohio to South Dakota and Kansas.

From the Corn Belt comes most of this country's annual eight billion plus bushels of corn, which is enough, says one mathematically minded urbanite, to bury all of Manhattan Island sixteen feet deep. About 50 percent of this corn is used as livestock feed. Another 25 percent is exported; 10 percent—optimally within five minutes of picking—ends up on the American dinner table, and the rest, in more or less adulterated forms, passes into an enormous number of peripheral corn products. Among these are corn syrup, corn starch, corn oil, and corn meal, the last once recommended (rapidly eaten in quantity) to prevent internal damage due to inadvertently swallowed fishbones. More versatile medicinally was corn whiskey, which was used to treat colds, coughs, consumption, toothache, rheumatism, and arthritis—all ailments that must have been rampant, since American whiskey consumption in the period 1790–1840 has been estimated at an annual five gallons a head.

Today processed corn in various forms figures in cardboard, charcoal briquettes, crayons, fireworks, wallpaper, aspirin, chewing gum, pancake mix, shoe polish, ketchup, marshmallows, instant tea, mayonnaise, surgical dressings, and soap. The cobs make a dandy fuel for smoking hams, and, carved, form the bowls of corncob pipes—a craft that went commercial in 1869 in Missouri. Cobs have also functioned as bottle stoppers, tool handles, hair curlers, and, sliced crosswise, checkers; creative housewives in nineteenth-century Missouri turned them into corncob jelly. The husks stuffed early American mattresses—Abraham Lincoln was born on a bed of cornhusks and bearskins in a cabin somewhere south of Hodgenville, Kentucky—or were woven into horse collars, and a process for turning cornhusks into paper was patented in 1802. Even the stalks have their uses: in the 1930s a group of researchers at Iowa State College came up with a concrete-like building material made from pulverized cornstalks, reported to be as hard as stone and stronger than wood. It was called maizolith. In the seventeenth-century Massachusetts Bay Colony, corn was accepted as legal tender, at least in transactions that didn't specify cold cash or beaver pelts.

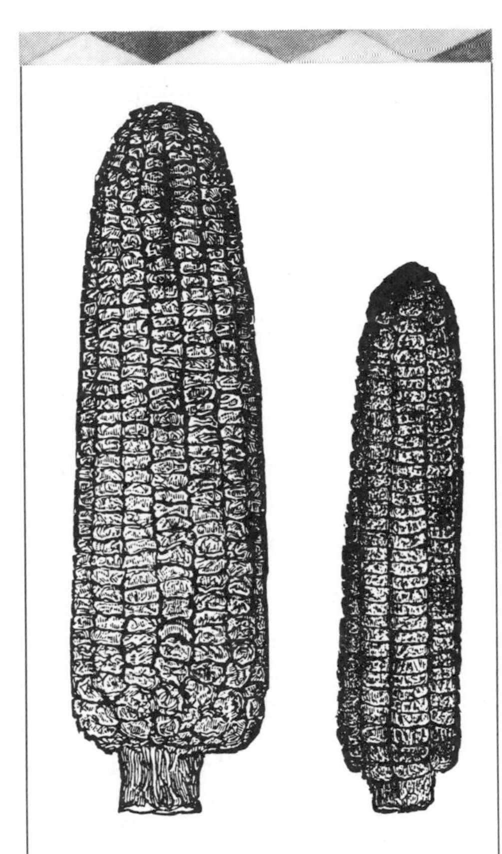

In the seventeenth-century Massachusetts Bay Colony, corn was accepted as legal tender, at least in transactions that didn't specify cold cash or beaver pelts.

For all the ingenious uses to which corn plants have been put over the past several centuries, botanists are cagily vague about where our national plant came from in the first place. Chances are that corn-on-the-cob started life as a not-too-promising breed of wild grass somewhere in the highlands of southern Mexico. The precise ancestry of corn is a botanical mystery, and speculation concerning it has reached such a pitch in scientific circles that the debates are known as the Corn Wars.

One prominent hypothesis holds that corn is descended directly from teosinte, an annual wild grass native to Mexico, Guatemala, and Honduras. Unlike modern corn, teosinte puts out multiple stalks from the base of the plant—a characteristic known as tillering—and bears triangular kernels enveloped in rock-hard seed cases. The kernels, six to ten in all, are arranged in a scrawny row—called the female spike—which is the teosinte equivalent of the present-day ear. The spike, however, shatters at maturity to release the seeds, while the modern ear, described by unsympathetic scientists as a "biological monstrosity," possesses no mechanism for successful seed dissemination. A fallen ear of corn lies there like a dud bomb, eventually sprouting a collection of infant seedlings so closely packed that their intense competition for water, food, and sunlight kills them off.

Proponents of the teosinte hypothesis—notably Nobel-prize-winning geneticist George W. Beadle (originally from Nebraska)—argue that modern corn and annual teosinte are very closely related genetically. Both possess ten pairs of almost identical chromosomes, and the plants interbreed freely to produce feistily fertile hybrids. Furthermore, Beadle and cohorts feel that only two mutations would have been necessary to set teosinte on the right road to corn: the first, a conversion from a shattering to a non-shattering spike; the second, a switch from a hard to a soft seed case—and eventually, as in modern corn, to naked kernels, which means no seed case at all. The hard seed cases render teosinte next to inedible. Beadle hypothesizes the prehistoric eaters probably popped it, by throwing the seeds on hot rocks or sand, or perhaps ground it between stones. Dr. Beadle states dauntingly that "an energetic person motivated by sufficient hunger" could probably scrape together enough meal by these primitive methods to feed a small family, and he went on to prove his point by living on teosinte for four days. A soft seed case simplified matters, however, even for the energetic and motivated, and such a corn exists today. Called pod or tunicate corn, this breed develops cobs that look as though they are upholstered in straw. Each kernel is covered by a soft, easily removable envelope known to botanists as a glume and to lesser folk as chaff. Pod corns, which dominate early archaeological sites, seem to have been the earliest cultivated corns.

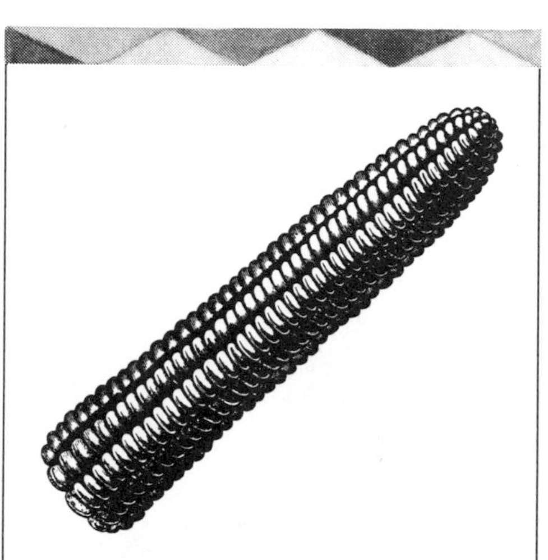

A fallen ear of corn lies there like a dud bomb, eventually sprouting a collection of infant seedlings so closely packed that the intense competition for water, food, and sunlight kills them off.

The opposing corn camp, led by Paul C. Mangelsdorf (originally from Kansas), holds that Beadle has put the cart before the horse: pod corn, according to Mangelsdorf and company, chanced to cross about four thousand years ago with a related wild grass (*Tripsacum*) to yield both annual teosinte and modern corn. Pollen-structure studies, however, put *Tripsacum* out of the running as an ancestral candidate, and recently the hypothesis has been modified, replacing the demoted

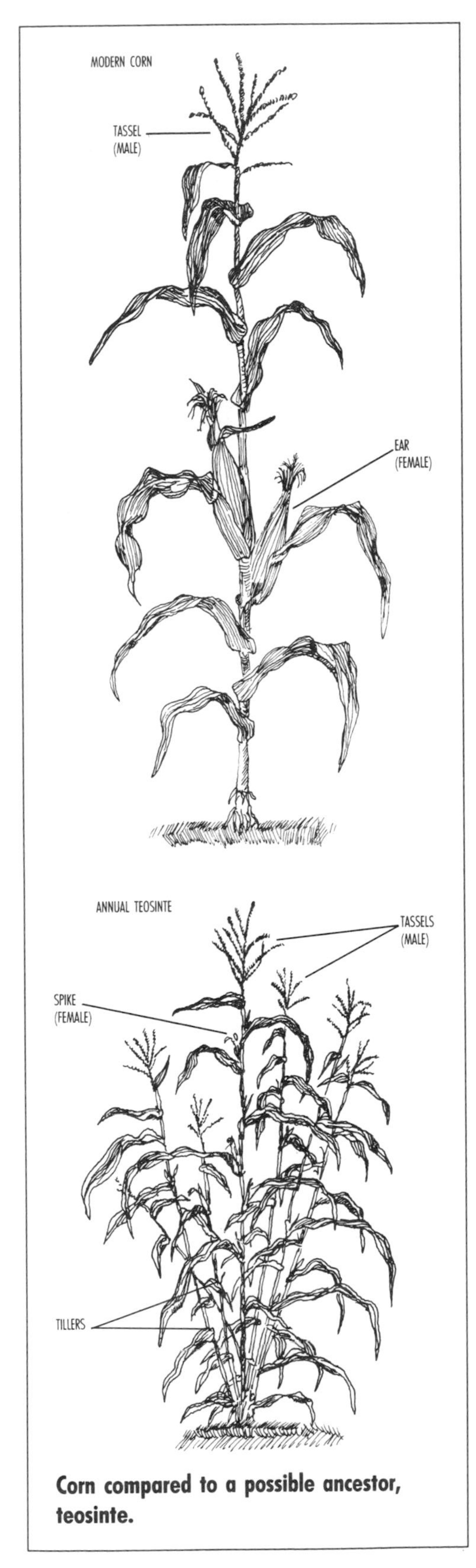

Corn compared to a possible ancestor, teosinte.

Tripsacum corn parent with perennial teosinte. Perennial teosinte (*Zea diploperennis*), the latest addition to the corn ancestry squabble, was discovered in 1979 by a sharp-eyed graduate student named Raphael Guzmán in Jalisco, Mexico. Until then, the only perennial teosinte known was a tetraploid species, carrying four sets of chromosomes instead of the more usual two. Such a wealth of genetic information is a reproductive handicap: the tetraploid teosinte can be coaxed into crossing with modern corn, but the (triploid) progeny are sterile. *Z. diploperennis*, which features stronger stalks and larger fleshier roots than its annual relative, is a diploid, possessing the same chromosome number as modern corn. Experimental crosses—initially performed in Mangelsdorf's back yard in Chapel Hill, North Carolina—between *Z. diploperennis* and Palomero Toluqueño, a primitive Mexican popcorn, yielded fertile offspring, some resembling annual teosinte, others with features suggestive of modern corn.

The very latest in corn ancestry hypotheses, however, comes from Hugh Iltis, at the University of Wisconsin in Madison. Iltis believes that corn evolved from annual teosinte by way of a bizarre vegetable sex change, referred to in the scientific literature as the "catastrophic sexual transmutation." Both modern corn and teosinte are monoecious, bearing separate male and female flowers on the same plant. In corn, the male flower—the tassel—develops at the very top of the cornstalk, where it manufactures a spectacular thirty to sixty million pollen grains per plant each growing season. These miniscule pollen grains—only 1/250 of an inch in diameter—are carried by the wind to land on the styles of neighboring female flowers. The styles, or silks—the stuff Tom Sawyer used to smoke—each extend down to a corn ovary which, when fertilized, develops into a full-fledged kernel. The many-kerneled cobs develop at the terminal ends of very short lateral branches sticking out from the main stem, or stalk.

Unlike modern corn, teosinte is very highly tillered—that is, along with a central stem, it produces many equivalently sized lateral branches from the same base. On the end of each develops a male tassel. Female "ears," or spikes, grow laterally off these lateral branches. Thus, in teosinte a male tassel sits at the terminal end of the lateral branches; in modern corn, lateral branches terminate in a female cob. Iltis, after mulling this over, concluded that a possible mechanism for the teosinte-into-corn transformation might be sexual flip-flop: the changing of the terminal (teosinte) male tassel into the terminal (corn) female ear. Such a switch, accompanied by shortening of the lateral branches of the present-day stubs and suppression of lateral teosinte-type female ears, essentially equals modern corn.

The Mexicans, who should know, call teosinte "the mother of maize," and in the 1950s a group of corn scientists jumped the gun by dubbing a professional journal *Teocentli* in honor of their supposed vegetable founder. Still, Drs. Beadle and Magelsdorf, now in their eighties, show no signs of compromise, and the corn feud, unresolved, is moving into second and third generations. The native Americans hold simpler views: the Navajos claim that corn came from a magical turkey hen, which dropped a providential ear of corn (blue) while en route to the morning star; and the Rhode Island Indians say the original corn was dropped by a crow.

To date, the earliest domesticated corn—a pathetic production compared to the juicy cobs cultivated today—has been excavated from archaeological sites in Mexico and the American Southwest. Dr. Herbert W. Dick, a Harvard anthropologist, unearthed primitive one- to two-inch cobs from beneath six feet of accumulated prehistoric garbage in the Bat Cave of New Mexico in 1948. Dated to thirty-five hundred years ago by radiocarbon analysis, the Bat Cave corn held the corn age record until the 1960s, when samples of a punier and more primitive five thousand-year-old corn—the cobs the size of pencil erasers—were discovered in the Tehuacán Valley of Mexico by Richard MacNeish of Boston University. Such authentic early corn specimens replaced an earlier corn find, a puzzlingly modern-looking ear of petrified maize that turned up in the early twentieth century in a curio shop in Cuzco, Peru. The ear, believed to be thousands of years old, was dubbed *Zea antiqua*. It was donated to the Smithsonian Museum, where, in the 1930s, botanists gingerly cut it open and found it to be made of pottery clay.

The styles, or silks—the stuff Tom Sawyer used to smoke—each extend down to a corn ovary which, when fertilized, develops into a full-fledged kernel.

Columbus saw vast cornfields in the West Indies, and likely brought the first samples home on his second transatlantic voyage in 1493, along with an assortment of parrots. Within two generations of discovery, the terrific new crop had spread to Africa, India, Tibet, and China, where the emperor, with a sharp eye for the main chance, was already taxing it. Along with the seed itself, Columbus brought back the native word for the grain—*mahiz*—which survives as the correct common name for American corn, *maize*. Much early confusion relative to the historical cultivation of maize stems from the European use of the word *corn* in a generic sense to mean kernel, as in *peppercorn* and *corned beef*, which was prepared with kernels (corns) of salt. *Corn* was also used to refer to whatever the dominant grain of the country happened to be. Hence, *corn* in England meant wheat; in Scotland oats; and in the Bible, where Ruth stood about in it miserably, it may have meant barley. The all-purpose term promptly expanded to include the new American grain. By 1542, when herbalist Leonhard Fuchs published the first known maize illustration, he called it Turkish corn and claimed it came from Asia. John Hariot, who published the first English description of maize in his "Briefe and True Report of the New Found Land in Virginia" (1588), skirted the issue by referring to the new grain by its Indian name, *Pagatowr*. Ten years later, John Gerard, who either never read Hariot or rejected him on the grounds of unpronounceability, called it Turkey or "Guinney" wheat, and thought it fine for the "barbarous Indians" but certainly not fit for upstanding Englishmen ("of hard and evill digestion, a more convenient food for swine than for man"). The Turks, who denied all connection with it, called it Egyptian corn; the Egyptians called it Syrian corn; the Germans threw up their hands and called it *Welschkorn*, which means strange grain. Linnaeus, who assigned the scientific name in 1737, dubbed it *Zea mays*—*mays* a spelling variant of the original *maize*, and *Zea*, meaning "I live," from the Greeks, who lightheartedly assigned the term to a number of different plants, among them true wheat.

The ear, believed to be thousands of years old, was dubbed *Zea antiqua*. It was donated to the Smithsonian Museum, where, in the 1930s, botanists gingerly cut it open and found it to be made of pottery clay.

Prehistoric farmers, determinedly selecting for larger, meatier ears, had developed an impressive two to three hundred different corn breeds by the time Columbus arrived, including all of the major classes of corn grown today: pop, flint, flour, dent, and sweet. The five classes differ primarily in the make-up of the corn endosperm, the food storage organ of the kernel. Stored food ranges from the tooth-cracking hard starch of pop and flint corns to the soft sugar of sweet corn. Most corn kernels contain around 70 percent starch, stored in tiny granules that form gradually, layer by layer, around a central nucleus. In popcorn, these granules are embedded in a tough matrix of protein. The kernel pops because, when heated, the internal water reaches the boiling point, vaporizes, and rapidly expands in volume, upping the pressure on both this protein matrix and the kernel's outer hull. When the pressure becomes insupportable, the matrix gives way and the kernel explodes—literally everting, or turning itself inside out—from which explosive property comes the scientific name *Zea mays* ssp. *everta*. Proper popping depends on water content: optimum is 13–14.5 percent, and Orville Redenbacher, who boasts nearly 100 percent popability for his gourmet popcorn, claims a precise moisture level of 13.25 percent for each kernel. Most popcorn, popped, swells thirty-five to thirty-eight times its starting size, a property processors refer to as the expansion ratio. Gourmet popcorn has a mammoth 44-fold expansion ratio and is described romantically by Redenbacher as looking like a cumulus cloud. (He also mentions that it makes a good substitute for Styrofoam chips as a packing material.)

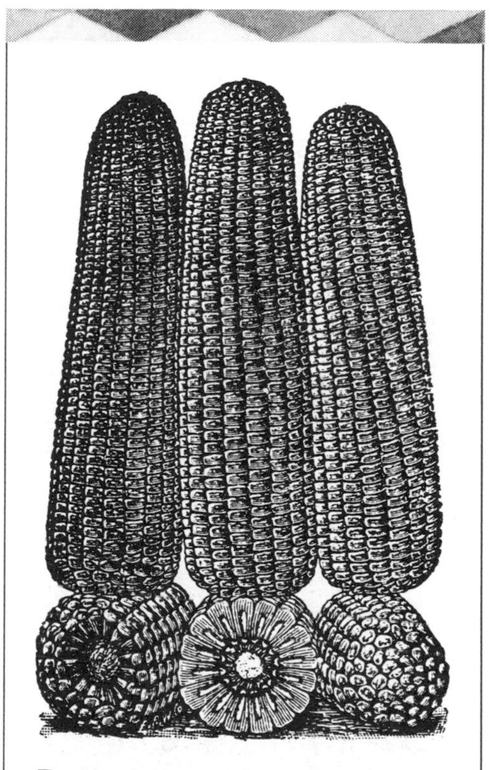

Prehistoric farmers, determinedly selecting for larger, meatier ears, had developed an impressive two to three hundred different corn breeds by the time Columbus arrived, including all of the major classes of corn grown today: pop, flint, flour, dent, and sweet.

The white crispy innards, of which Americans collectively crunch through 9.7 billion quarts per year, consist mostly of cooked starch. The colonists came across their first popcorn at the famous first Thanksgiving, when Quadequina, brother of Massasoit, showed up with several pre-popped bushels in a deerskin bag. Colonial popcorn was eaten for breakfast in New England, with milk and maple sugar, and the Pennsylvania Dutch, who also gave us Christmas cookies and apple strudel, cooked up chicken corn soups with popcorn floating on the top to give them extra oomph. By the mid-nineteenth century, popcorn was no longer a mealtime staple, but was still such a popular munchie that vendors were hawking it on the city streets. Popcorn balls, coated with molasses, caramel, or honey, moved in as candy treats in the 1870s, and soon ranked right up there with the old-time favorites, vinegar candy and saltwater taffy. Two entire buildings at the Philadelphia Exposition of 1876 were devoted to the selling of popcorn, which was washed down with Arctic Soda Water, elegantly dispensed from colored marble soda-water fountains with silver fittings. By the turn of the century, popcorn was bought and sold from gorgeously decorated steam-powered popcorn wagons designed by a baker named Charles Cretors. The wagons, complete with shiny red paint, beveled glass, and brass edgings, are collector's items today. Cretors went on to patent the electric corn popper in 1916, thus paving the way for the present-day popcorn industry. In the late 1970s, popcorn graduated abruptly from the standard salt-and-butter fare of movie theaters to a specialty snack. In its latest incarnation, luridly tinted in colors ranging from hot-pink to lavender, popcorn is flavored with pineapple and pepperoni, raspberry and root beer, chocolate, coffee, cantaloupe, pistachio, and butter rum.

Popcorn is an extra-hard form of flint corn, the somewhat larger high-protein, hard-starch corn that predominated in the northern United States and Canada centuries before the Europeans arrived. Today it's usually colored flint corn that people use to decorate their front doors in autumn, calling it "Indian corn." Early native growers purposefully selected and bred their crops for color, producing, by the time the settlers arrived, red, blue, black, yellow, white, and multicolored corns. Intrigued colonists continued these breeding practices, and eventually grew ears of every conceivable hue, including purple, maroon, amber, chocolate-brown, lemon-yellow, copper, and orange.

Early native growers purposefully selected and bred their crops for color, producing by the time the settlers arrived, red, blue, black, yellow, white, and multicolored corns.

Popular fashion today narrow-mindedly limits most eating corns to yellow or white—which shades, plus orange, result from the deposition of xanthophyll and carotene pigments in the endosperm of the kernel. Gaudier colors reside in the outer pericarp or in the aleurone layer, the one- to two-cell-thick, nutrient-packed sheet wedged between pericarp and endosperm. An aleurone-based blue accounts for the blue speckles—

the "moldly look," says one corn historian—in the very best corn tortillas, which, in the American Southwest, are made from blue corn. Kernel color in corn is cumulative, the final effect the result of pericarp pigmentation over aleurone layer over endosperm. Thus a blue aleurone layer topping a yellow endosperm may yield a pea-green corn. In Kculli, an ancient near-black corn used by the Peruvian Indians to make dye and colored beer, a dark red pericarp tops a purple or brown aleurone and a white endosperm.

Aleurone and pericarp may also be variegated: spotted, striped, or streaked in contrasting color combinations. Dr. Mangelsdorf proposes that this variegation in corn kernels is analogous to the variegation seen in "broken" tulips, a piebald or mottled color effect—quite gorgeous—generally attributed to virus infection. The appearance of colorful splotches in corn, Mangelsdorf hypothesizes, is controlled by a genetic modulator locus, which sporadically inhibits kernel pigmentation in the same manner that the artistic tulip virus controls petal color. Patriotically prominent among variegated corns is a red, white, and blue patterned novelty resembling the Stars and Stripes, developed by corn expert Walton Galinat.

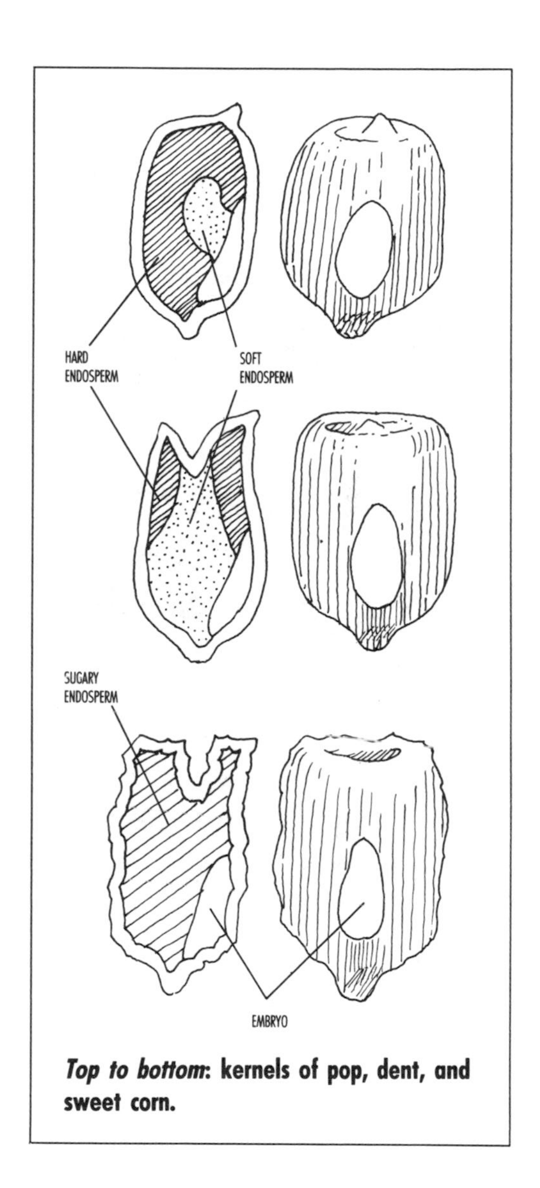

***Top to bottom*: kernels of pop, dent, and sweet corn.**

Another Indian specialty was flour corn—also brightly colored—which differs from its flinty relative by a mutation at a single locus on chromosome 2. Seldom grown today, flour corn featured soft, low-protein endosperm, easily ground into meal. Most common in the fields of the modern Corn Belt is dent corn, which contains a mix of hard and soft starch, the soft concentrated at the crown of the mature kernel. When dried, this soft starch shrinks, forming the characteristic dent. Hard and soft starches are made up of two different kinds of starch molecules. The hard stuff consists of straight chains of linearly linked sugars and is scientifically known as amylose; soft starch, composed of branched chains, is known as amylopectin. An unusual Asian breed of corn, introduced to this country in the early 1900s from China, contains only branched starch. This is known as waxy corn, from the slick wax-like appearance of the cut grain, and is the corn equivalent of the "glutinous" character in sticky rice.

Sweet corn was grown in prehistoric Mexico and Peru, and by many of the North American Indian tribes. Most sources state that the pioneers didn't manage to get their hands on it until 1779, when Richard Bagnal, an officer in General John Sullivan's destructive expedition against the Iroquois, nabbed a sample from a cornfield on the Susquehanna River in western New York. Sweet corn is higher in sugar than in starch, and produces irresistibly delicious translucent wrinkled kernels analogous to the equally wrinkled seed of high-sugar peas. Sweet corn is rendered so by a gene mutation known as *shrunken* that interrupts the ordinary conversion of sugar to dextrin (a middle-sized polysaccharide) to long-chain starch. It appears to be a fairly common mutation as such things go, occurring at the rate of two sweet mutants per million gametes, which means, according to Paul Mangelsdorf, one newly appearing sweet mutant in every seventeen acres of planted corn.

Classically, sugar starts going downhill the minute the corn is picked—hence the traditional adage that you can stroll out to the cornpatch as slowly as you please, but you'd better run like the devil back to the house.

Even more sugary is supersweet corn, governed by a mutation identified in 1950 by researcher J. R. Laughman at the University of Illinois. Called *shrunken 2*, this mutation halts the sugar-to-starch conversion at an earlier point. Sweetness increased yet again in the 1960s, when another Illinois scientist, A. M. Rhodes, discovered a positively sinful three-way cross that he called *sugary enhanced* (*se*) corn. Marketed *se* corns include cultivars in the Everlasting Heritage series—Kandy Korn EH, for example—and the sweet bicolor Burgundy Delight. The *se* mutation enables corn to maintain its sugar content much longer after harvest than is usual. Classically, sugar starts going downhill the minute the corn is picked—hence the traditional adage that you can stroll out to the cornpatch as slowly as you please, but you'd better run like the devil back to the house. Mark Twain took that one further, recommending that the boiling cooking pot be taken right out into the garden. Actually, that's an unnecessarily drastic step; corn experts today grant pickers a twenty-minute grace period before the deterioration begins.

The sugar in corn provides a fine food source for yeasts, and corn from prehistoric times has been used to make beer and wine, and later, whiskey. In the starchier types of corn, the starch molecules must be broken down into their component sugars before the yeasts can go to work. The ancient Peruvians solved the starch problem by chewing the ground corn, thus breaking down the corn starch with the human salivary enzyme, amylase. The resultant mash was then fermented to yield an alcoholic beverage called *chicha*, which was not only enthusiastically drunk, but also sprinkled in the fields at corn-planting time to ensure a good harvest. Post-sprinkling, the imperial Inca himself started the sowing by breaking the soil with a golden pickaxe.

A related enzymatic process, similar in theory if not in practice, is used today to convert corn starch to corn syrup. The enzymes in this case are obtained from *Aspergillus oryzae*, a cultured mold, and act to reduce the starch molecules to progressively shorter sugar units. In Japan, *A. oryzae* is used to convert rice starch to sugar in preparation for making *sake*. Alternatively, the corn can be malted, or allowed to germinate for several days until the kernel's own enzymes break the stored starch down into sugar to provide energy for the sprouting (but doomed) seedling. The early colonists, deprived of English barley, made beer of malted Indian corn—blue corn was the best, according to local lore—and occasionally, when desperate, of dried persimmons, pumpkins, or Jerusalem artichokes. They also made brandy out of crushed cornstalks and wine from fermented cobs.

More high-powered corn liquors appeared with the establishment of the Scotch-Irish whiskey distilleries in eighteenth-century Pennsylvania—though the first known whiskey-maker in North America was a Dutchman, William Kieft of New Amsterdam, who turned out his potent beverages on Staten Island in the 1640s. The Pennsylvania distillers were heavily taxed by the financially strapped federal government after the Revolutionary War, and the tax cut into their profits to such an extent that they banded together in the Whiskey Rebellion of 1794. The rebellion failed—George Washington unsympathetically suppressed it with fifteen thousand federal troops—and many of the rebels packed up their stills and moved to Kentucky, where they began making bourbon.

Corn whiskey-making, however, continued: Thomas Jefferson, who really preferred wine, started

CORN BEER WITHOUT YEAST

From Dr. Chase's Recipes, *or,* Information for Everybody, *1869*

♦ Cold water 5 gals.; sound nice corn 1 qt.; molasses 2 qts.; put all into a keg of this size; shake well, and in 2 or 3 days a fermentation will have been brought on as nicely as with yeast. Keep it bunged tight.

It may be flavoured with oils of spruce or lemon, if desired, by pouring on to the oils one or two quarts of the water, boiling hot. The corn will last five or six makings. If it gets too sour add more molasses and water in the same proportions. It is cheap, healthy, and no bother with yeast.

making it at Monticello in 1813; John C. Fremont carried it, ostensibly as an antifreeze for his surveying instrument, during his western explorations of the 1840s; and the South mourned its absence during the Civil War, when the North captured the Tennessee copper mines and cut off the supply of metal for kettles and condensing tubes. Among the best of existing corn whiskey stories appears in the after-action report of a Yankee major, Isaac Lynde, who in July 1862 was commanding a garrison of soldiers en route to Fort Stanton in New Mexico. His men turned out to have canteens full of whiskey, snitched from medical stores in the dispensary, and, being in no state to either stand and fight or run away, were promptly captured by patrolling Texans, who were sober. The humiliated Major Lynde would doubtless have thrown in his lot with Edward Enfield, whose 1866 "Treatise on Indian Corn" condemned whiskey, stating that corn "like some other best gifts of the Deity, [has] been perverted to base and injurious uses."

His men turned out to have canteens full of whiskey, snitched from medical stores in the dispensary, and, being in no state to either stand and fight or run away, were promptly captured by patrolling Texans, who were sober.

When not drunk in "spirituous liquors," corn was consumed in fritters, hoecakes, pone, hominy, and hasty pudding—the last so ubiquitous that colonial poet Joel Barlow wrote a lengthy poem in honor of it. The Indians made their own brand of hasty pudding, flavored with dried blueberries and grasshoppers. Benjamin Franklin, who liked it for breakfast, flavored his with honey and nutmeg, and the French, who adopted it as a political gesture during the American Revolution, flavored theirs with cognac. Corn was also eaten roasted green as the familiar corn-on-the-cob—a delightful pastime that arises, according to corn scientist Walton Galinat, from "man's instinctive desire to eat directly with his bare hands." The accompanying table manners made European visitors shudder.

For all the appeal of corn cuisine, corn, nutritionally, has its definite drawbacks. None of the cereal grains is a complete protein source and corn is no exception. It is deficient in both the essential amino acids lysine and tryptophan, and not too forthcoming with the vitamin niacin, which in corn kernels is tightly bound to another molecule and thus is unavailable to hopeful eaters. A lack of niacin—the name was coined to avoid the nasty cigarette-smoking connotations of the scientifically correct nicotinic acid—leads to a deficiency disease called pellagra in human beings and blacktongue in dogs.

The cause of pellagra, a particular problem in the poverty-stricken South, was a puzzle for decades. One hypothesis held that the sufferers were weak by nature and acquired the disease in an unmentionable manner from sheep. Finally, in 1915, Joseph Goldberger, a Public Health Service physician, performed a series of ominous-sounding nutritional studies at orphan asylums, the results of which pinpointed pellagra as a dietary deficiency disease. It was non-infectious, linked to "low-grade starchy diets," and curable by supplementing the diet with meat, milk, and eggs. The amino acid tryptophan is a precursor of niacin, which means that in the absence of ready-made niacin, the body can manufacture its own, provided it is supplied with proteins rich in tryptophan. Corn in this case administers a nutritional double whammy: its niacin exists in unusable form and it lacks tryoptophan.

None of the cereal grains is a complete protein source and corn is no exception.

Once this was known, scientists began to wonder why prehistoric societies based on corn, such as the Mayas and Aztecs, were not riddled with pellagra. The reason lies with their corn-processing technique. Before grinding their corn into meal, primitive societies subjected the kernels to an alkali treatment—boiling in wood ashes or lime—that both helped in the removal of the hulls and improved the overall nutritional quality of the cornmeal. The nutritional benefits are upped, perversely, by cutting down the availability of zein, the principal storage protein in corn, and the worst of the corn proteins in terms of its rockbottom content of lysine and tryptophan. This in turn increases the relative availability of these scarce amino acids. Alkali-treated corn provides the eater with nearly three times the lysine of untreated corn, plus converts the bound niacin to a useful absorbable form. Corn tortillas are thus a better protein source than corn-on-the-cob. Hominy, from the Algonquian *rockahominie* or *tackhummin*, is similarly alkali-treated: the grains are soaked in lye extracted from wood ashes, then washed to remove the hulls, and pounded. The result, hominy grits, is a standard addition to Southern breakfasts, though

Northerners tend to compare it scornfully to bird seed.

In the 1960s, three researchers at Purdue University discovered a variant of corn with twice the lysine content of the normal plant. The variant, designated *opaque-2*, generated high hopes, but was soon found to have substantially lowered yields and increased susceptibility to disease. The latest approach to the nutritional failures of corn involves genetic engineering. Molecular biolgoists hope to manipulate the genes coding for the zein protein—which accounts for over half the protein in the average corn kernel—to increase its normally skimpy allotments of lysine and tryptophan.

Before geneticists started fiddling around with the DNA helix, corn improvement was pursued by more conventional means. Among the first to notice the phenomenon of cross-fertilization in corn was Cotton Mather, who, peering into his neighbor's garden in 1716, noticed that yellow corn planted downwind from red and blue ended up with multicolored ears. Most early corn breeding was pursued in this happy-go-lucky fashion: different varieties of corn planted next to each other crosspollinated to yield new and occasionally better offspring. It was in this offhand manner that Robert Reid produced his famous Yellow Dent, by interspersing a dent corn called Gordon Hopkins Red with Little Yellow, an early flint corn, on his farm south of Peoria, Illinois. Reid's Yellow Dent, cited as "the most significant stride in corn production since prehistoric times," won a prize at the Chicago World's Fair of 1893—belatedly, since by then it had been America's favorite corn for decades. Such free and easy crosspollination is viewed with a cold eye by home gardeners who want to grow more than one kind of corn in the same small vegetable patch. To avoid undesirable sexual intermixing, gardening pro Nancy Bubel suggests sternly separating corn types with rows of sunflowers to act as towering pollen traps.

The first deliberate studies of corn crosses were performed by no less an authority than Charles Darwin, who published the results in 1876 under the title *The Effects of Cross and Self-Fertilisation in the Vegetable Kingdom*. It made a good deal less of a splash than the inflammatory *On the Origin of Species by Means of Natural Selection, or the Preservation of Favoured Races in the Struggle for Life* (1859), which sold out, all 1,250 copies, on the day of issue, but it nonetheless made a few useful points. Foremost among these was the observation that the offspring of crosses between related strains were weaker and less productive than those of crosses between unrelated parents—or conversely, that the progeny of unrelated parents are healthier and more vigorous. This phenomenon is known as heterosis or hybrid vigor, and, on the human scale, is why kings occasionally marry commoners, with an eye to beefing up the royal bloodline.

An American disciple of Darwin, William James Beal at the Michigan Agricultural College (now Michigan State University), performed the first controlled crosses in corn. Beal planted his corn patch with two different varieties of corn, then put one over on Mother Nature by detasseling—emasculating—corn number one, thus making certain that any ears would result only from cross-fertilization with pollen from the tassels of corn number two. The hybrid offspring of this carefully directed interaction, right on cue, were healthier, happier, and higher-yielding than the wimpish progeny of closely related parents.

The man usually credited with the introduction of vigorous hybrids to the commercial cornfield is George Harrison Shull, who in photographs looks very like the current fashion in scientists, sporting a beard and wire-rimmed spectacles. Shull carried out his hybrid experiments around the turn of the century at Cold Spring Harbor, Long Island. He planted his first corn patch with an eye to tourists, designing a vegetable garden demonstration of the principles of Mendelian genetics, but soon abandoned such scientific frivolities in favor of serious hybrid research. Shull's major contribution was the development of pure inbred parental strains of corn, by repeated generations of self-fertilization. Once his collection of pure lines was established, Shull tried a cross, and in one fell swoop obtained vigorous, high-yielding, first-generation hybrids. Hybrid corn seed went on the market in the 1930s, and high-yielding hybrids now occupy the vast majority of American cornfields.

Unfortunately, farmers have paid for their in-

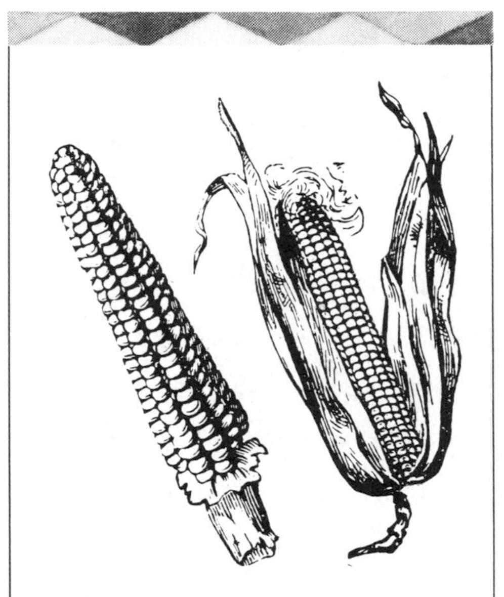

Among the first to notice the phenomenon of cross-fertilization in corn was Cotton Mather, who, peering into his neighbor's garden in 1716, noticed that yellow corn planted downwind from red and blue ended up with multicolored ears.

creased production figures with decreased genetic diversity—which means that in the 1970s, over 70 percent of American corn acreage was planted with just six varieties of corn. Such a degree of specialization was asking for trouble, and in 1970, when a new strain of southern leaf blight fungus appeared on the scene, identically susceptible corn plants dropped like flies across the nation. Sensitivity to the attacking fungus was correlated to possession of the otherwise useful Texas (T) cytoplasmic male sterility factor, a genetic element that functions like a vegetable vasectomy, eliminating the need for painstakingly detasseling the plants to prevent self-fertilization. The result was an overall 15 percent loss of the corn crop, and up to 50 percent in some unlucky states. In light of this painful lesson, research efforts now are directed toward incorporating a greater range of characteristics into existing corn varieties. Even more than a vacuum, it seems, Nature abhors uniformity.

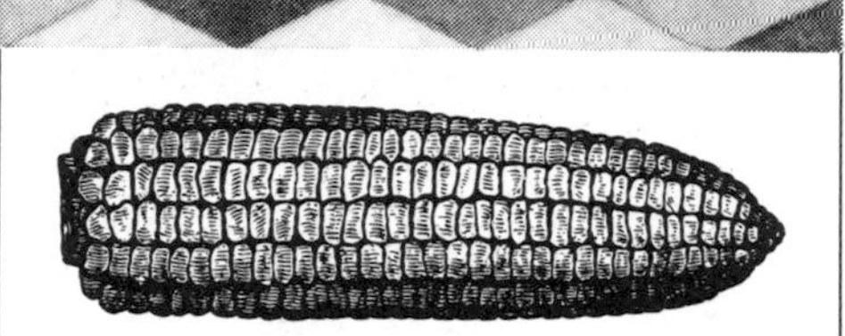

In the 1970s, over 70 percent of American corn acreage was planted with just six varieties of corn. Such a degree of specialization was asking for trouble, and in 1970 when a new strain of southern leaf blight fungus appeared on the scene, identically susceptible corn plants dropped like flies across the nation.

Uniformity, however, was an obsession among corn breeders in the early years of the twentieth century, when a craze known as the Corn Show swept the country. Farmers competed to produce the biggest, the best, and the most mathematically perfect ear of corn, the hopefuls often judged in theatrical Corn Palaces especially constructed for this purpose. Only one such Palace still survives, in Mitchel, South Dakota, built in the days when the citizens of Mitchel decided to outshine the glamorous celebrations of the Grain Palace in the neighboring town of Plankinton. Mitchel's Corn Palace, built in 1921, boasts turrets, towers, onion domes, and an immense arched Arthurian gateway, all seasonally decked out in elaborate mosaics made from two thousand bushels of variously colored corns. In its heyday, the Corn Palace played host to the rich and famous: William Jennings Bryan and Teddy Roosevelt both spoke there, and, in 1904, John Philip Sousa treated the assembled corn competitors to a series of twelve concerts, at a total cost to the town of seven thousand dollars. (The concerts nearly fell through at the last minute, when Sousa, suspicious, demanded cash in hand before letting his band get off the train.)

The Corn Show was on its way out by the 1920s, largely due to the commonsensical urgings of agriculturalist Henry Wallace. Wallace, never one to choose a house for its paint or a book for its cover, pointed out that the purpose of 85 percent of the American corn crop was to feed animals, not win beauty contests. "Looks," announced Wallace, "mean nothing to a hog." Wallace's own homely hybrids, created with the hog in mind, were astonishingly high yielders. Among them was a variety called Copper Cross, descendant of a gore-colored colonial corn named Bloody Butcher, with which Wallace went on to found the now gargantuan Pioneer Hi-Bred seed company. Wallace himself became Secretary of Agriculture, and later Vice President, under Franklin Delano Roosevelt, and ran ineffectively for president in 1948, raking in 2.8 percent of the national vote. His corn company did better, and Pioneer these days accounts for 35 percent of annual seed corn sales, distributing some six hundred billion hybrid kernels across the Corn Belt each year. At an estimated 1:800 return on each kernel, that's a lot of corn.

Cornflakes were so popular that by the early 1900s some forty-four cereal companies—some of them in tents—were in business in and around Battle Creek.

Corn in America is often still eaten three meals a day, though the morning dollop of cornmeal mush has been replaced in the twentieth century by the bowl of cornflakes. The name most commonly associated with this breakfast treat is Kellogg, of whom there were two, the brothers John H. and Will K. John H. Kellogg was a doctor who, fresh out of medical school in 1876, took over the directorship of the Western Health Reform Institute at Battle Creek, Michigan. Under his rule, the Institute became famous for its specially designed vegetarian diets: the skinny were fed twenty-six meals a day and made to stay in bed with sandbags on their stomachs; the hypertensive were served nothing but grapes; and everybody was encouraged to gnaw zwieback, for the health of gums and teeth. Within the first year, Kellogg was plying patients with his first toasted cereal creation—a mix of smashed-up biscuits of oats, wheat, and cornmeal known as Granose. It took him until 1895 to come up with the first flake cereal, a wheat preparation made by rolling partially cooked whole grains out flat, then toasting until crisp and dry. (The idea, said Dr. Kellogg, had come to him in a dream.) The public turned up its collective nose at wheatflakes, but cornflakes, brought out several years later, were a stunning success. Cornflakes were so popular that by the early 1900s some forty-four cereal companies—some of them in tents—were in business in and around Battle Creek. Prominent among cereal promoters was C. W. Post, a suspender salesman turned health-foods manufacturer, whose personal brand of cornflakes went on the market in 1906 under the name Elijah's Manna. Both name and carton—which showed an assortment of heavenly ravens dropping cornflakes into the hands of a hungry prophet—were considered blasphemous by clergymen, and turned out to be downright illegal in Britain, where it was forbidden to register Biblical names for commercial purposes. Post reissued his flakes in 1908 as Post Toasties.

Kellogg's cornflakes (flavored with barley malt) were the foundation of the enormous Battle Creek Toasted Corn Flake Company, established in 1906 by Dr. Kellogg's entrepreneurial brother Will. John and Will seldom saw eye to eye over the proposed direction of the breakfast food industry, and their differences of opinion landed them in court for twelve years of legal tussling over who was legally entitled to the use of the Kellogg name. Will K. won (with a few limited privileges to John H.), which is why it's Will K.'s signature that gives the seal of approval to the modern cornflakes box. Both John and Will, nourished on cornflakes, lived well into their nineties. Will K. died in 1951; his tombstone, suitable for the prototypic morning man, is a sundial bearing a bronze robin pulling up a bronze worm. It could almost be an emblem for gardeners everywhere.

SEED SOURCES

One of the high points of the bleak and debt-ridden months after Christmas is the arrival of the seed catalogs. Perusal of these, in the intemperate zones, is for many the sole saving grace of February; and even the pro-snow enjoy an occasional bask by the fire with the latest in garden bulletins, preferably illustrated in full and luscious dew-drenched color.

These days catalog devotees are in clover. Hundreds of such publications are now available, ranging in quality from the one-page basement-mimeographed flier to the glossily professional botanical booklet.

A number of these mail-order seed sources are listed below, categorized, somewhat fuzzily, by vegetable emphasis. The *traditional suppliers* here are the general practitioners of the garden world, providing an all-purpose selection of seeds for the average vegetable patch. More specialized are the *heirloom seed suppliers*, committed to the propagation of historical, usually open-pollinated, vegetable varieties, many of which are in danger of disappearing from the modern garden. *Gourmet seed distributors* pander to the culinary-minded, providing exceptionally flavorful, often imported, vegetable varieties. (Think *cornichon*, not dill pickle.) And finally the *seed specialists*, a mixed bag—categorizers of seed companies often run into these difficulties—committed to a single vegetable or to a restricted assortment of unusual vegetables.

TRADITIONAL SUPPLIERS

W. Atlee Burpee Company
300 Park Avenue
Warminster, Pennsylvania 18974

The Old Faithful of mail-order seed houses, the W. Atlee Burpee Company has been around since 1876 when the 18-year-old founder issued his first four-page catalog, a poultry-dominated publication with a picture of ducks on the cover. Among the vast assortment of flower and vegetable seeds offered are the famous Big Boy tomato ("a garden favorite since 1949"), the Snowbird Marigold, the first named variety of pure-white marigold, and Green Ice lettuce, a ruffled looseleaf distinguished by its possession of the first awarded plant patent. Burpee also sells fruit and nut trees, berry bushes, assorted garden tools, and Italian honeybees. Catalog free.

Henry Field's Seed and Nursery Company
Shenandoah, Iowa 51602

Founded in 1892, Henry Field's

catalog today includes among its flower, herb, and vegetable assortment eight kinds of okra, in red, white, and green, Sweet Potato squash, Green-Striped cushaw ("a really different pumpkin"), a hominy corn collection (Reid's Yellow Dent, White Hickory King, and Trucker's Delight), and, diplomatically, both Vidalia and Walla Walla onions. Catalog free.

Gurney Seed and Nursery Company
Yankton, South Dakota 57079

Gurney's, in the seed business since the close of the Civil War, sends out an oversized catalog crammed with planting tips, photographs of tickled customers, and usual and unusual seeds. Among these are Beautiful beans, an (almost) gasless cream-and-maroon heirloom, the German Giant radish ("big as baseballs"), the Sweet Yellow Dumpling potato, a yellow-fleshed heirloom from Nebraska, and an assortment of old-fashioned sweet corns, including Golden Bantam, Country Gentleman, and Stowell's Evergreen. Catalog free.

Harris Seeds
Moreton Farm
3670 Buffalo Road
Rochester, New York 14624

Joseph Harris's "Select List of Field, Garden and Flower Seeds" first appeared in 1879. Today the "Select List" is a 96-page color catalog of vegetable and flower seeds, scattered with homey snapshots of toddlers perched on pumpkins and grandmothers hefting cantaloupes. Harris also offers an inflatable Great Horned Owl, guaranteed to scare birds, rabbits, and squirrels away from the crops (three for 13.50). Catalog free.

J. L. Hudson, Seedsman
P.O. Box 1058
Redwood City, California 94064

J. L. Hudson offers a sizeable collection of "seeds of the rarer sort," in a catalog that reads and looks like a small dictionary, the plants listed in alphabetical order by scientific name. Among the rarities are Japanese pumpkins, Bloody Butcher and Pencil Cob corns, and heirloom Sulphur beans. For the more ambitious, the company offers the silk cotton tree, which bears fruit filled with silky fibers suitable for stuffing pillows, the night-blooming cereus, a 10- to 40-foot cactus with immense nocturnal white flowers, and the vegetable hummingbird, a tropical tree whose flowers, boiled, are said to taste like mushrooms. Catalog $1.00.

Johnny's Selected Seeds
Foss Hill Road
Albion, Maine 04910

A delightful, attractive, and informational catalog, Johnny's offers vegetable and flower seeds, grains, cover crops, garden supplies, books, and a 100 percent cotton Johnny's Selected Seeds T-shirt. Among the selected vegetable seeds are Case Knife beans, "one of the oldest known green beans in America," Tongue of Fire shell beans, whose ivory pods are streaked with flame-red, Maple Arrow soybeans, a yellow variety from Canada, miniature purple onions (they turn pale-pink when pickled), and old-fashioned Midnight Snack sweet corn, whose kernels are white at the eating stage, blue-black when dried at maturity. Catalog free.

J. W. Jung Seed Company
Randolph, Wisconsin 53957

J. W. Jung started his seed business on the family farm in 1907, and today, at 100 years old, he still lives just next door, close enough to keep a weather eye on the crops. The Jung Seed Company mails out over a million catalogs annually, featuring a substantial selection of vegetable and flower seeds. Among those available are three-inch-diameter miniature pumpkins, Banana Blue squash, three varieties of kohlrabi, and an array of extraordinarily sweet hybrid corns. Also from Jung: an old-fashioned cherry pitter that fits any standard mason jar and a twelve-room purple martin house. Catalog free.

Earl May Seed and Nursery Company
Shenandoah, Iowa 51603

The Earl May Seed and Nursery Company—"Home Planting Headquarters for the Nation Since 1919"—still subscribes to their founder's praiseworthy, though slightly ungrammatical, philosophy: "I'm not happy if your not happy." Included on their cheerful vegetable list are golden beets, a pure white onion dubbed Miss Society ("Eat onions and go to the party. So mild scarcely noticed by your friends."), Mammoth Jumbo peanuts ("produce a crop as far north as Wisconsin"), and seedless watermelons. Earl May also offers a selection of home and garden gadgets, including a corn desilking brush, originally developed as a surgeon's scrub brush, a press for making radish roses, and two kinds of pea-shellers, hand-cranked and electric. Catalog free.

Mellinger's, Inc.
2310 West South Range Road
North Lima, Ohio 44452-9731

Mellinger's—"The Garden Catalog for Year-Round Country Living"—tucks their six black-and-white pages of vegetable seeds at the back of the catalog, behind flowers, trees, shrubs, grasses, cover crops, and garden supplies. They also sell herb plants and seeds, including seven kinds of mint (applemint, spearmint, orange mint, pineapple mint, curled or bergamot mint, and a tall variegated spearmint called Emerald 'n' Gold). A source for duck whirligigs, inflatable scarecrows, and pink flamingos. Catalog free.

Park Seed Company
Cokesbury Road
Greenwood, South Carolina 29647-0001

Founded in 1868 by George W. Park, whose photograph, showing impressive handlebar moustaches, still appears in the modern catalog, the Park Seed Company remains very much in the family. Park offers a complete selection of vegetable and flower seeds, cunningly packaged for freshness in laminated gold foil packets. Included are sugar beets, pear-, pineapple-, and fruit punch-flavored melons, Tom Thumb butterhead lettuce ("midget tennis ball size heads"), purple bell peppers, and Homestead Hybrid potato seed ("the first hybird Potato from seed"), reported to have two- to four-fold higher yields than the famous Explorer potato seed. Catalog free.

Pinetree Garden Seeds
Route 100
New Gloucester, Maine 04260

One of the thrills of Pinetree Garden Seeds is their price: offered in "packages of reasonable size" (minimum number of seeds per packet is listed for each vegetable), most Pinetree packets cost less than fifty cents and a few only cost a quarter. A quick whip through their price list is like time travel. Offered in these bargain bundles are flower, herb, and vegetable seeds, plus a recently added specialty section of "Vegetable Favorites from Around the World," featuring radicchio, purple artichokes, daikon radishes, cardoons, celeriac, and cutting celery. Catalog free.

Stokes Seeds, Inc.
Box 548
Buffalo, New York 14240

Stokes has been selling "Quality Garden Seeds" since 1881. The modern catalog, a no-frills black-and-white production, is jam-packed with vegetable, herb, and flower seeds, both open-pollinated and hybrid varieties. Many seeds are chemically treated, but untreated equivalents are often available for organic gardeners. Stokes vegetables include low-acid tomatoes, naked-seeded pumpkins, "bitterfree" pickling cucumbers, and leafless peas. Catalog free.

Vermont Bean Seed Company
Garden Lane
Fair Haven, Vermont 05743

In a delightful and enchantingly illustrated catalog, the Vermont Bean Seed Company offers not only an excellent selection of bean seeds, but everything else for the well-equipped garden from Artichoke to Watermelon, varieties both old-fashioned and brand-new. Vegetable possibilities include Wren's Egg, Vermont Cranberry, Yellow Eye, Jacobs Cattle, and Black Turtle Soup beans, edible burdock, Marble White cantaloupes (16 percent sugar), White Wonder cucumbers, Sweet Dumpling, a green-and-white-striped edible gourd, and heirloom Gilfeather turnips. There's also a company T-shirt, with a silk-screened kidney bean on the chest. Catalog free.

HEIRLOOM SUPPLIERS

Abundant Life Seed Foundation
P.O. Box 772
Port Townsend, Washington 98368

Founded by agriculturalist Forest Shomer in 1972, the Abundant Life Seed Foundation is a nonprofit corporation dedicated to the "heirloom seed renaissance." Their catalog, appropriately printed in green, offers a large selection of untreated, open-pollinated seeds: vegetables, annual and perennial herbs, flowers, and grains. Among the vegetables are Appaloosa and Buckskin beans, Oxheart carrots, Deertongue lettuce, Milk pumpkins, an old-fashioned light-tan variety, Mandan Bride, a five-color flour corn, and Bearpaw popcorn. Catalog $1.00.

Blue Corn Nursery
Route 10, Box 87 R
Santa Fe, New Mexico 87501

A one-page price list of ancient southwestern corns, red, white, and multicolored, as well as blue. "If your corn does very well," the Nursery urges generously, "share the wealth by giving seeds to your friends, relatives, and neighbors everywhere . . ." Price list free.

Bountiful Gardens
Ecology Action
5798 Ridgewood Road
Willits, California 95490

Ecology Action, a nonprofit organization founded in 1972, is heavily involved in organic biointensive food-raising research, practices described in the popular *How to Grow More Vegetables (Than You Ever Thought Possible on Less Land Than You Can Imagine)* by member John Jeavons. Their mail-order catalog offers untreated, open-pollinated heirloom vegetable, herb, and flower seeds ("The best seeds are handed down from one generation to the next as heirlooms, just like rare silver or china"), along with grains, "green manure" crops, garden supplies, and a large selection of books. Catalog free.

Good Seed Company
P.O. Box 702
Tonasket, Washington 98855

Good Seed ("Modern and Heirloom Seed for Northern and Mountain Gardens") sells only untreated, open-pollinated seeds, under the premise that "one thing this planet does not need is yet another merchant offering the same lineup of proprietary national hybrid seeds." Among the comprehensive selection of vegetables, herbs, flowers, and cover crops are Ice Cream watermelons, Hopi Black Dye sunflowers, blue-speckled tepary beans, purple-and-white Cherokee Princess corn, New England Pie pumpkins, and Egyptian Walking onions. Also available are a number of seed collections, including a Pioneer Vegetable Garden. Catalog $1.00.

Heirloom Gardens
P.O. Box 138
Guerneville, California 95446

A seed list rather than a catalog, of "culinary, historic and rare plant seeds," among them wild oats, dyer's chamomile, spoon gourds, soapwort, lemon catnip, and Big Red, an early colonial corn with red or red-and-white-striped kernels. Seed list free.

Native Seeds/SEARCH
2950 West New York Drive
Tucson, Arizona 85745

A nonprofit organization dedicated to the preservation of native crops of the southwest United States and northwest Mexico. Among their selection are tepary beans, annual and perennial teosinte, indigo, Apache Brown Striped sunflowers, Canteen gourds, and 75 different native corns, including Navajo Robin's Egg (speckled blue, white, and red), Hopi Chin Mark (red-and-white-striped), and Chapalote (brown). They also sell several Native American cookbooks, featuring recipes such as Pinto Beans with Watermelon Seeds, Yucca Pie, and Pinon Nut Soup. Catalog $2.00.

Seeds Blüm
Idaho City Stage
Boise, Idaho 83706

Seeds Blüm (pronounced *bloom*) specializes in "Heirloom Seeds and Other Garden Gems": vegetables, herbs, and flowers, all open-pollinated. Their clever and conversational catalog features, among many others, Green and Purple Yardlong beans, four kinds of Jerusalem artichokes, Purple Calabash and Persimmon tomatoes, blue potatoes, Hopi Blue Flint and Bloody Butcher corns, Moon and Stars watermelons, skirret, rat-tail radishes, jelly melons, and cardoons. There's also a Trading Post for seed swappers and, this year, a tempting recipe for red carnation marmalade. Catalog $2.00.

GOURMET SUPPLIERS

The Cook's Garden
P.O. Box 65
Londonderry, Vermont 05148

The Cook's Garden offers "Seeds, Plants and Produce for the Serious Cook"—particularly if that serious cook is a creative salad-maker. At last count, the catalog carried 54 different lettuce varieties, plus a wide selection of unusual salad plants: chicory, chervil, cutting celery, cress, and three kinds of mesclun, a tasty mix of leafy salad greens. Also Planet carrots (nearly round), white and purple sprouting broccoli, Principe

Borghese tomatoes ("grown in Italy for drying"), and Vermont maple syrup. Catalog $1.00.

Le Jardin du Gourmet
West Danville, Vermont 05873

A small catalog, but a large assortment of herbs and French vegetable seeds, plus Portuguese cabbages, African pumpkins, and German beer garden radishes, which grow as big as turnips (you slice them and dip the slices in sugar). Also offered are an array of gourmet delicacies, including snails (and snail forks), paté, caviar, lobster paste, and six kinds of mustard, and a selection of books, featuring *Dowsing for Everyone* ($8.95) in honor of the American Society of Dowsers, which is located in Danville, Vermont. Catalog $0.50.

Le Marché Seeds International
P.O. Box 566
Dixon, California 95620

Le Marché, state co-owners Georgeanne Brennan and Charlotte Glenn, was "founded on the premise that home gardeners are seeking vegetable varieties whose primary attribute is flavor." Their catalog of vegetable and herb seeds, temptingly dotted with recipes, also makes for flavorful reading. Among their offerings are Violet cauliflowers (serve raw, since they turn green when cooked), Black popcorn, cornichon cucumbers, Santa Claus casabas (will keep until Christmas), Alpine strawberries, and White Cheesequake squash, a cheesebox-shaped pumpkin first listed for sale in 1824 by New York seedsman Grant Thorburn. Catalog $2.00.

Nichols Garden Nursery
1190 North Pacific Highway
Albany, Oregon 97321

Nichols Garden Nursery specializes in international gourmet vegetable seeds, culinary and ornamental herbs, and garden flowers, with the avowed purpose of bringing people "closer to nature through gardening." Their large and thoroughly enjoyable catalog listing includes Elephant garlic ("bulbs weighing up to one pound or more each"), Orangetti squash, a vivid orange spaghetti squash, gynoecious (all-female) greenhouse cucumbers, husk tomatoes, daikon radishes, black salsify, and papaya pumpkins. Also offered are a selection of garden and kitchen supplies, books, herbal teas, and beer- and wine-making aids (champagne corks: 10/$1.45). Catalog free.

Shepherd's Garden Seeds
7389 West Zayante Road
Felton, California 95018

Shepherd's Garden Seeds was established "to offer American gardeners the opportunity of growing choice European vegetable varieties." Among the vegetable prizes listed in this mouthwateringly descriptive catalog are Italian and scented basils, Dutch baby beets, Japanese hybrid broccolis, Rollinson's Telegraph cucumber, a British "burpless" variety ("perfect for cucumber sandwiches"), miniature pumpkins, and Yellow Pear tomatoes. Catalog $1.00.

SPECIALISTS AND ODDITIES

Gleckler's Seedmen
Metamora, Ohio 43540

The seedmen at Gleckler's are purveyors of "unusual seed specialties," both imported and homegrown. Among the specialties listed in their six-page green-printed catalog are Pink cress, a dry land cress imported from India, Pink Bride eggplant, a pink-and-white-striped hybrid, Long Handled Dipper or Baton gourd (record length: 72 inches), Chinese edible-seeded watermelon, Evergreen and Pink Grapefruit tomatoes, and the Cob melon, a cream-fleshed muskmelon in which the seeds form on a corn-like cob and thus can be quickly removed. Gleckler's also offers, for the older gardener, Natural Beauty Creme, an organic skin rejuvenator devised by a Mr. Clarence H. Prager of Oklahoma. Catalog free.

Grace's Gardens
10 Bay Street
Westport, Connecticut 06880

Grace's Gardens issues the "World's Most Unusual Seed Catalog"—and certainly one of the smallest, two double-sided pages and only 21 seed varieties, nine of them tomatoes. Among these are the Giant Belgium tomato (for wine), the Square tomato (for paste), the Hollow tomato (for stuffing), and the all-white White Beauty tomato (for startling the neighbors). Also Italian edible gourds and three-foot-long Kyoto cucumbers. Seed list free.

High Altitude Gardens
P.O. Box 4238
Ketchum, Idaho 83340

High Altitude Gardens, located at a chilly 6,000 feet, concentrates "on the challenge of gardening in harsh climates with its inherent, short-growing season." Their attractive catalog offers a sizeable assortment of open-pollinated vegetable, herb, flower, and grass seeds, selected for high performance on the heights. Among this feisty crew are Windsor fava beans ("frost-hardy jewels"), Blue Max Savoy cabbages ("will survive temperatures as low as 10 degrees F"), Polarvee hybrid corn (67 days to maturity), and Stupice tomatoes (52 days to maturity, "the earliest tomato we have discovered"). Catalog $2.00.

The Pepper Gal
Dorothy L. Van Vleck
10536 119th Avenue North
Largo, Florida 33543

The Pepper Gal—a hobby pepper grower gone professional—offers pepper aficionados a list of 248 varieties, ornamental, sweet, and hot. Particularly recommended for those who like their chili three-alarm are Cowhorn and Tennessee Firecracker. Price list free.

Siberia Seeds
Box 2026
Sweetgrass, Montana 59484

Ron Driskill's tomato company came into being in 1983, with a sole spectacular product: the Siberia tomato, 48 days to maturity and capable of setting fruit at 38 degrees F. Siberia Seeds has since added another dozen cold-hardy tomatoes to its seed list, among them the Glacier, a Canadian heirloom, orginally from Sweden, and Landry's Russian, an impressive high-producer brought to Canada "many years ago" by Russian immigrants. Seed list $0.50 or send self-addressed stamped envelope.

The Tomato Seed Company, Inc.
P.O. Box 323
Metuchen, New Jersey 08840

The catalog offers over 300 varieties of heirloom and modern tomatoes, grouped by size: Huge, Extra Large, Large, Medium, Small and Cherry, Pear and Plum. Among these are the Crimson Cushion (ribbed red), the Watermelon Beefsteak (a pink-fruited heirloom), the Goldie (a gold-colored heirloom dating to 1800), the White Beauty (an all-white high-sugar tomato), and the Doublerich (extra high in vitamins A and C), plus the Red Currant tomato (pea-sized fruits), the Tree tomato, and green and purple tomatillos.

BIBLIOGRAPHY

Adams, John F. *Guerilla Gardening.* New York: Conrad-McCann, 1983.

Alberts, Robert C. "The Good Provider." *American Heritage,* February 1972, pp. 26–47.

Aldrich, Nelson. "A Social History of Corn." *Blair and Ketchum's Country Journal,* September 1983, pp. 46–52.

The American Heritage Cookbook and Illustrated History of American Eating and Drinking. New York: American Heritage Publishing Co., 1964.

Anderson, Edgar. *Plants, Man, and Life.* Berkeley: University of California Press, 1969.

Andrews, Jean. *Peppers: The Domesticated Capsicums.* Austin: University of Texas Press, 1984.

Barkas, Janet. *The Vegetable Passion.* New York: Charles Scribner's Sons, 1975.

Barker, Lewis M., ed. *The Psychobiology of Human Food Selection.* Westport, Conn.: AVI Publishing Co., 1982.

Barton, Kenneth A. and Winston J. Brill. "Prospects in Plant Genetic Engineering." *Science* 222:671–676.

Beadle, George W. "The Ancestry of Corn." *Scientific American* 242:112–119.

Bennett, Jennifer. "A Tomato Blossom for All Seasons." *Horticulture,* March 1983, pp. 53–59.

Booth, Sally Smith. *Hung, Strung & Potted: A History of Eating Habits in Colonial America.* New York: Clarkson N. Potter, 1971.

Borlaug, Norman E. "Contributions of Conventional Plant Breeding to Food Production." *Science* 219:689–693.

Brewster, Letitia and Michael F. Jacobsen. *The Changing American Diet.* Washington, D.C: Center for Science in the Public Interest, 1981.

Bridges, Bill. *The Great American Chili Book.* New York: Rawson, Wade Publishers, 1981.

Brothwell, Don and Patricia. *Food in Antiquity.* New York: Frederick A. Praeger, 1969.

Brown, John Hull. *Early American Beverages.* New York: Bonanza Books, 1966.

Brown, William L., "Hybrid Vim and Vigor." *Science 84,* November 1984, pp. 77–78.

Bryant, John A. "Genetic Vectors for Plants: Regulator Expression of a Foreign Gene in Cells of Tobacco." *Trends in Biotechnology* 2:108–9.

Bubel, Nancy. "Supersweet Corn." *Blair and Ketchum's Country Journal*, June 1982, pp. 24–25.

Bubel, Nancy. "The Elegant Eggplant." *Horticulture*, February 1986, pp. 26–28.

Burton, W. G. *Post-harvest Physiology of Food Crops*. Essex, England: Longman Group, 1982.

Carson, Gerald. "Cornflake Crusade." *American Heritage*, June 1957, pp. 66–85.

Chaleff, R. S. "Isolation of Agronomically Useful Mutants from Plant Cell Cultures." *Science* 219:676–682.

Christopher, Thomas. "The Great Tomato Quest." *Horticulture*, March 1985, pp. 19–22.

Clarkson, Rosetta E. *The Golden Age of Herbs and Herbalists*. 1940. Reprint. New York: Dover Publications, 1972.

Collins, Elizabeth P. "Seed Coatings." *Garden*, March/April 1981, pp. 14–17.

Cook, Jack. "The Explorer Potato." *Horticulture*, July 1983, pp. 10–15.

Cornog, Mary W. *Growing and Cooking Potatoes*. Dublin, N.H.: Yankee Publishing, 1981.

Dark, Sandra. "Which Shell Beans to Grow for Drying?" *Blair and Ketchum's Country Journal*, May 1982, pp. 52–56.

de Forest, Elizabeth Kellam. *The Gardens and Grounds at Mount Vernon*. Mount Vernon, Va.: The Mount Vernon Ladies' Association of the Union, 1982.

Dodge, Bertha. *Plants That Changed the World*. Boston: Little, and Brown Co., 1959.

Donovan, Mary, Amy Hatrack, Frances Mills, and Elizabeth Shull. *The Thirteen Colonies Cookbook*. New York: Praeger Publishers, 1975.

Doty, Walter L. and A. Cort Sinnes. San Francisco: *All About Tomatoes*. Ortho Books, 1981.

Drummond, J. C. and Anne Wilbraham. *The Englishman's Food*. London: Jonathan Cape, 1939.

DuBose, Fred. *The Total Tomato*. New York: Harper and Row, 1985.

Fairbrother, Nan. *Men and Gardens*. New York: Alfred A. Knopf, 1956.

Faust, Joan Lee. *The New York Times Book of Vegetable Gardening*. New York: Quadrangle, The New York Times Book Co., 1975.

Favreti, Rudy F. and Gordon P. DeWolf. *Colonial Gardens*. Barre, Mass.: Barre Publishers, 1972.

Fielder, Mildred. *Plant Medicine and Folklore*. Tulsa, Okla.: Winchester Press, 1975.

Fincher, Jack. "Tailored Genes." *Horticulture*, April 1984, pp. 50–57.

Fisher, M. F. K. *The Art of Eating*. New York: Vintage Books, 1976.

Friend, J. and M. C. Rhodes, eds. *Recent Advances in the Biochemistry of Fruits and Vegetables*. New York: Academic Press, 1981.

Garfield, Eugene. "From Tonic to Psoriasis: Stalking Celery's Secrets." *Current Comments*, May 6, 1985, pp. 3–12.

Gould, Stephen J. "A Short Way to Corn." In *The Flamingo's Smile*, pp. 360–373. New York: W. W. Norton and Co., 1985.

Grant, Susan. *Beauty and the Beast: The Coevolution of Plants and Animals*. New York: Charles Scribner's Sons, 1984.

Greenberg, Emmanuel. "The Great Popcorn Explosion." *Playboy*, March 1984, pp. 88, 194–195.

Greene, Bert. *Greene on Greens*. New York: Workman Publishing, 1984.

Hardeman, Nicholas P. *Shucks, Shocks, and Hominy Blocks*. Baton Rouge: Lousiana State University Press, 1981.

Harlan, Jack R. *Crops and Man*. Madison, Wis.: American Society of Agronomy and Crop Science Society of America, 1975.

Haughton, Claire Shaver. *Green Immigrants*. New York: Harcourt Brace Jovanovich, 1978.

Hedrick, U. P. *A History of Horticulture in America to 1860*. New York: Oxford University Press, 1950.

Heiser, Charles B., Jr. *Seed to Civilization*. San Francisco: W. H. Freeman and Co., 1981.

Heiser, Charles B., Jr. *Of Plants and People*. Norman: University of Oklahoma Press, 1985.

Hendrickson, Robert. *The Great American Tomato Book*. Garden City, N.Y.: Doubleday and Co., 1977.

Herman, Judith and Marguerite Shalett Herman. *The Cornucopia*. New York: Harper and Row, 1973.

Hibben, Frank C. *Digging Up America*. New York: Hill and Wang, 1960.

Hickman, Peggy. *A Jane Austen Household Book*. North Pomfret, Vt.: David and Charles, 1977.

Hooker, Richard J. *Food and Drink in America*. New York: The Bobbs-Merrill Co., 1981.

Horry, Harriott Pinckney. *A Colonial Plantation Cookbook: The Receipt Book of HPH, 1770*. Columbia: University of South Carolina Press, 1984.

Hulme, A. C., ed. *The Biochemistry of Fruits and Their Products*. New York: Academic Press, 1971.

Iltis, Hugh H. "From Teosinte to Maize: The Catastrophic Sexual Transmutation." *Science* 222:886–94.

Jabs, Carolyn. *The Heirloom Gardener*. San Francisco: Sierra Club Books, 1984.

Kahn, E. J., Jr. *The Staffs of Life*. Boston: Little, Brown and Co., 1985.

Katz, S. H., M. L. Hediger, and L. A. Valleroy. "Traditional Maize Processing Techniques in the New World." *Science* 184:765–773.

Khudairi, A. Karim. "The Ripening of Tomatoes." *American Scientist* 60:696–707.

Kimball, Michael. "On Farting." *Coevolution Quarterly*, Summer 1982, pp. 80–85.

Kraft, Ken and Pat. *The Best of American Gardening: Two Centuries of Fertile Ideas*. New York: Walker and Co., 1975.

Leighton, Ann. *Early American Gardens "For Meate or Medicine."* Boston: Houghton Mifflin Co., 1971.

Leighton, Ann. *American Gardens in the Eighteenth Century.* Boston: Houghton Mifflin Co., 1976.

Lovelock, Yann. *The Vegetable Book: An Unnatural History.* New York: St. Martin's Press, 1972.

Lyte, Charles. *The Kitchen Garden.* Somerset, England: The Oxford Illustrated Press, 1984.

Mangelsdorf, Paul C. *Corn: Its Origin, Evolution and Improvement.* Cambridge, Mass.: Harvard University Press, The Belknap Press, 1974.

Mangelsdorf, Paul C. "The Origin of Corn." *Scientific American,* August 1986, pp. 80–87.

Mangelsdorf, Paul C. and Robert G. Reeves. *The Origin of Indian Corn and Its Relatives.* Texas Agricultural Experiment Station Bulletin, no. 574, May 1939.

Maugh, Thomas H. "It's Nothing to Cry About. . . ." *Science* 204:293.

McDonald, Lucile. *Garden Sass: The Story of Vegetables.* N.P.: Thomas Nelson, 1971.

McGee, Harold. *On Food and Cooking.* New York: Charles Scribner's Sons, 1984.

McKendry, Maxime. *Seven Hundred Years of English Cooking.* New York: Exeter Books, 1973.

Mennell, Stephen. *All Manners of Food: Eating and Taste in England and France from the Middle Ages to the Present.* New York: Basil Blackwell, 1985.

Messing, Joachim. "The Manipulation of Zein Genes to Improve the Nutritional Value of Corn." *Trends in Biotechnology* 1:54–59.

Mintz, Sidney W. *Sweetness and Power: The Place of Sugar in Modern History.* New York: Viking Penguin, 1985.

Phipps, Frances. *Colonial Kitchens, Their Furnishings and Their Gardens.* New York: Hawthorn Books, 1972.

Pizer, Vernon. *Eat the Grapes Downward: An Uninhibited Romp Through the Surprising World of Food.* New York: Dodd, Mead and Co., 1983.

Proulx, E. A. "Some Like Them Hot." *Horticulture,* January 1985, pp. 46–54.

Proulx, E. Annie. *The Fine Art of Salad Gardening.* Emmaus, Penn.: Rodale Press, 1985.

Quinn, Vernon. *Leaves: Their Place in Life and Legend.* New York: Frederick A. Stokes Co., 1937.

Quinn, Vernon. *Vegetables in the Garden and Their Legends.* New York: J.B. Lippincott Co., 1942.

Ralston, Nancy C. and Marynor Jordan. *Garden Way's Zucchini Cookbook.* Pownal, Vt.: Garden Way Publishing, 1977.

Raymond, Dick and Jan. *The Gardens for All Book of Potatoes.* Burlington, Vt.: Gardens for All, 1980.

Redenbacher, Orville. *Orville Redenbacher's Popcorn Book.* New York: St. Martin's Press, 1984.

Revel, Jean-François. *Culture and Cuisine*. Garden City, N.Y.: Doubleday and Co., 1982.

Rick, Charles M. "The Tomato." *Scientific American*, August 1978, pp. 76–87.

Riotte, Louise. *Carrots Love Tomatoes*. Pownal, Vt.: Garden Way Publishing, 1975.

Robertson, Laurel, Carol Flinders, Bronwen Godfrey. *Laurel's Kitchen*. Petaluma, Calif.: Nilgiri Press, 1976.

Root, Waverley. *Food*. New York: Simon and Schuster, 1980.

Root, Waverley and Richard de Rochemont. *Eating in America*. New York: The Ecco Press, 1976.

Roueche, Berton. *The Orange Man and Other Narratives of Medical Detection*. Boston: Little, Brown and Co., 1965.

Rozin, Elizabeth and Paul Rozin. "Culinary Themes and Variations." *Natural History*, February 1981, pp. 6–14.

Russell, Howard S. *A Long, Deep Furrow: Three Centuries of Farming in New England*. Hancock, N.H.: University Press of New England, 1976.

Saeger, Elizabeth. *Gardens and Gardeners*. New York: Oxford University Press, 1984.

Sann, Paul. "The Time of the Green." In *Fads, Follies and Delusions of the American People*, pp. 133–135. New York: Bonanza Books, 1967.

Shapiro, Laura. *Perfection Salad: Women and Cooking at the Turn of the Century*. New York: Farrar, Strauss, and Giroux, 1986.

Shell, Ellen Ruppel. "Palate-Pleasing Through Art and Science." *Smithsonian*, May 1986, pp. 79–88.

Shepard, James F., Dennis Bidney, Tina Barsby, and Roger Kemble. "Genetic Transfer in Plants Through Interspecific Protoplast Fusion." *Science* 219:683–688.

Sifakis, Carl. *American Eccentrics*. New York: Facts on File Publications, 1984.

Simmonds, N. W. *Evolutions of Crop Plants*. Essex, England: Longman Group, 1976.

Simmons, Amelia. *The First American Cookbook*. A facsimile of *American Cookery*, 1796. New York: Oxford University Press, 1958.

Singer, Marilyn. *The Fanatic's Ecstatic Aromatic Guide to Onions, Garlic, Shallots and Leeks*. Englewood Cliffs, N.J.: Prentice-Hall, 1981.

Sokolov, Raymond. "Broad Bean Universe." *Natural History*, December 1984, pp. 84–86.

Spitler, Sue and Nan Hauser. *The Popcorn Lover's Book*. Chicago: Contemporary Books, 1983.

Stuart, David C. *The Kitchen Garden: A Historical Guide to Traditional Crops*. London: Robert Hale, 1984.

Sturtevant, E. Lewis. *Sturtevant's Notes on Edible Plants*. 1919. Reprint. *Sturtevant's Edible Plants of the World*. New York: Dover Publications, 1972.

Swain, Roger. *Earthly Pleasures*. New York: Charles Scribner's Sons, 1981.

Tarr, Yvonne Young. *The Squash Cookbook*. New York: Random House, 1978.

Trager, James. *The Food Book*. New York: Grossman Publisher, 1970.

Vilmorin-Andrieux et Cie. *The Vegetable Garden*. London: John Murray, 1885.

Wheaton, Barbara Ketcham. *Savoring the Past: The French Kitchen and Table from 1300 to 1789*. Philadelphia: The University of Pennsylvania Press, 1983.

Whiteside, Thomas. "Tomatoes." *The New Yorker*, January 24, 1977, pp. 36–61.

Wiencek, Henry. "House of Corn." *Americana*, September/ October 1983, pp. 110–112.

Williams, Susan. *Savory Suppers and Fashionable Feasts: Dining in Victorian America*. New York: Pantheon Books, 1985.

Zohary, D. and M. Hopf. "Domestication of Pulses in the Old World." *Science* 182:887–894.

INDEX

M

N

O

P

Q

R

S